M000203974

ATLANTA
BEER

ATLANTA

BEER

A HEADY HISTORY OF BREWING
IN THE HUB OF THE SOUTH

RON SMITH & MARY O. BOYLE

AMERICAN PALATE

Published by American Palate
A Division of The History Press
Charleston, SC 29403
www.historypress.net

Copyright © 2013 by Ron Smith and Mary O. Boyle
All rights reserved

First published 2013

ISBN 978.1.5402.3266.3

Library of Congress CIP data applied for.

Notice: The information in this book is true and complete to the best of our knowledge. It is offered without guarantee on the part of the author or The History Press. The author and The History Press disclaim all liability in connection with the use of this book.

All rights reserved. No part of this book may be reproduced or transmitted in any form whatsoever without prior written permission from the publisher except in the case of brief quotations embodied in critical articles and reviews.

CONTENTS

FOREWORD

I t's about time this book was written. Even though most of us think that the brewing world in Atlanta is just in its infancy (which, in a way, is true), it does have a legacy. It has stories of its own, and it is always nice to tell those stories. It has been fun hearing about the old breweries in Atlanta's past—where they were, the beers they brewed and how they brewed them. Certainly many of those tales are lost to time; I wish they weren't, as I would love to hear more of them.

In more recent years, especially in the past twenty or so, craft brewing has taken a foothold in Atlanta and has started to grow. The beer world is changing at such an incredible pace, and Atlantans are (re)discovering beer and brewing in larger numbers. Breweries are expanding, neighborhood brewpubs are opening their doors, great pubs and beer stores are thriving and people are enjoying an incredible variety of beers brewed right here in Atlanta. It is exciting to see and even more exciting to be a part of this story.

Even though the craft-brewing revolution might be in its early stages, this trade is growing and changing at an amazing pace. Because of this rapid rise, it is an excellent time to capture the past, to tell and to retell the stories so they are not forgotten. As Ron and I sat over a pint one evening and spoke about the "early days," it was fun simply to remember how it was twenty years ago, recalling the people, the breweries, the beers and the bars. Those are the stories that fill these pages.

It is also fun to dream of what is to come, just like we dreamed twenty years ago about what would happen next in Atlanta's brewing world. There are great things ahead in the local beer industry. I'm sure, some day in the future, there will be some sort of volume two to this book, filled with the next chapters. But for now, grab a beer, settle in and enjoy these tales: the stories of the beginnings and the present of beer in Atlanta.

CRAWFORD MORAN
Brew master
5 Seasons Westside and North, the Slice & Pint

ACKNOWLEDGEMENTS

Hundreds of hours of research went into the creation of this book. However, none of this research would have borne fruit if it were not for the aid of many brewers, establishment owners, collectors, research institutes, aficionados and all-around beer geeks that contributed knowledge, photos and humorous stories to this project.

COLLECTORS, HISTORIANS AND RESEARCH INSTITUTES

We wish to extend special thanks to Bill Baab, "the Bottle Man of Augusta"; Ken Jones, Atlantic breweriana collector extraordinaire; Tad Mitchell, breweriana collector and coowner of Six Feet Under Pub & Fish House; the Cuba Archives of the William Breman Jewish Heritage Museum (in particular Maureen MacLaughlin and the personal collection of Phyllis Levine); the James G. Kenan Research Center at the Atlanta History Center; Georgia State University Library, Special Collections (especially Ellen Johnston and Fray DeVore); the Georgia Archives; Georgians Against Closing State Archives for successfully lobbying to keep the Georgia Archives open; the Thomas Cooper and South Caroliniana Libraries of the University of South Carolina; the Digital Library of Georgia; and www.Fold3.com.

Acknowledgements

Bier Folken

Many people in the beer industry shared their personal knowledge through a series of interviews. A comprehensive list of these interviewees is in the bibliography. However, we wish to extend special thanks to John "JR" Roberts, brew master at Max Lager's Wood-Fired Grill & Brewery, for his repeated hospitality and history sharing; Crawford Moran, brew master at 5 Seasons Brewing, for his boundless help and graciousness; the entire crew at the Atlanta Brewing Company, aka Red Brick Brewing, for its overwhelming courtesy, exuberance for the book and brewery facts; George Lamb, formerly of Marthasville Brewing Company, for great details and humorous anecdotes; Gail Smith and Randy Dempsey of O'Dempsey's; Eddie Holley at Ale Yeah! for offering early assistance and contacts; Harry Hager, beer man extraordinaire, for contacts and local brewery scene information; Martha and George Karakos, former owner-operators of Atkins Park Tavern; and Alan Raines and Tryon Rosser of Hotoberfest for taking Ron on the "barrel run."

Others Deserving Mention

Special thanks go to Rose Dunning for providing custom artwork. Thank you to Judy Kuniansky for artfully shooting our authors' photo. We give a thanks to John and Christina "Red" Gagne (Brockett Pub House & Grill) for partnering with us over the years on beer dinners and a high five to Sam Covell for her careful review of the manuscript prior to submission. We extend gratitude to family and friends for providing encouragement and enthusiasm as we pursue our passions. Last, though chronologically first, a sincere thank-you goes to The History Press for extending the opportunity to write this book.

Cheers,
Ron Smith and Mary Boyle

INTRODUCTION

B eer, initially a food source, has for some time been a cultural mainstay and social lubricant. Beer in Atlanta's history is particularly interesting, as most historical writings on American beer focus on cities like Milwaukee, St. Louis and Cincinnati. Little has been written about the history of breweries in the Southeast. They have been mostly overlooked by authors who focus on fast cars, fast politics and the fast profits of Prohibition whiskey. Only a few diehard collectors have knowledge of the South's brewing heritage.

Beer is contextual. Many of us remember not only the flavors of our favorite beers but also the first places we enjoyed them, with whom and, sometimes, even the mood we were in at the time. The story of Atlanta beer is also contextual. It's a frontier story, an immigrant story, a saloon story, a religious and moral story, a story of decline and a tale that ends with an amazing comeback. This book covers a portion of the Greater Atlanta metropolitan statistical area—specifically, a subarea extending a nearly thirty-mile radius from downtown, with the addition of Cumming.

There are variations of a few words in the book, mainly due to historical usage. *Whisky* and *draught* are used in historical contexts, while *whiskey* and *draft* are used for modern times. Prohibition with an uppercase *P* is used to denote national Prohibition, while prohibition with a lowercase *p* is used for state and local anti-alcohol laws. The titles of brewers, business owners and coowners were taken directly from company information or personal communications and are a reflection of the research information, with no implication that one title is better than another.

Not all representatives of the beer industry in Atlanta could be interviewed for this book. The rapid growth of the industry and the time and coordination of schedules made it impossible to talk to everyone. Instead, this book is a carefully constructed cross-section of Atlanta's past and present beer culture. Our regrets are that we had limited time and limited words.

Writing this book was a learning process, both about brewing history (which Ron dearly loves) and about the people in the Atlanta beer scene. We already knew these folks were into beer. The extent of that passion was not fully revealed until time was spent interviewing brewers, bar and brewpub owners, distributors, homebrewers, breweriana collectors and craft beer and growler shop owners. Atlanta is on the verge of a beer renaissance along with the rest of the United States. With each sip, everyone writes the future.

CHAPTER ONE

FRONTIER BEER AND TAVERNS

If barley be wanting to make into malt,
We must be contented and think it no fault;
For we can make liquor to sweeten our lips,
Of pumpkins and parsnips, and walnut tree chips.
—*"Forefathers' Song," circa 1630*

FRONTIER BEER

Beer on the American frontier was dramatically different from beer of today—as different as the territory itself. The Georgia frontier was once the lands of the Mvskoke (Creek) and Cherokee Peoples, became the southwestern end of England's colonial American territory and, later, marked the southwestern border of the newly formed United States of America. Due to Georgia's distant location, supply lines into the area were slow and sporadic.

Brewing varied depending on the availability of fermentable items and seasonal temperatures. Before refrigeration (by natural ice or mechanical means), it was nearly impossible to keep beer for long in the summer heat, even in a cooler storage area. The beer would spoil due to poorly sealed containers and bacteria that thrived in the heat. In winter, not much was available to ferment, unless resourceful individuals had stored syrups, dried fruits or tubers or other starchy vegetable material.

Beer in the southern frontier was often made from a combination of corn, molasses or treacle mixed with one or more of the following: pumpkins, persimmons, maple or birch syrup, potatoes, sassafras roots, Jerusalem artichokes (a tuber from a type of sunflower) and even peas. If you were lucky, you might have a small amount of wheat, barley or the bran from a grain to add to your mash. Hops were an occasional item; in lieu of hops, gruit (grut) was used. In the Southeast United States, this herb mixture could be juniper berries or spruce tips, locust beans, various berries or medicinal and aromatic herbs.

In early Georgia, the most commonly documented beer is made from one or a combination of molasses, corn or persimmons. An intriguing persimmon beer recipe made from persimmon pone, water and locust beans is outlined in *Pigsfoot Jelly & Persimmon Beer*. (Pone is dried or fresh cake made of a base ingredient, water and seasoning. In the South, where corn has always been the staple crop, corn pone is most common.)

Corn was often used to make beer, especially during times of hardship, such as the Civil War and periods of alcohol prohibition. An 1800s recipe for corn beer was reprinted in the early Atlanta newspaper the *Southern Confederacy* in March 1862. It calls for a mash of molasses, corn and water, kept in a jug. By adding fresh water and more molasses, the same corn could be used for six months. Yum, yum.

As trade roads improved, Asian ginger became available. Ginger was first brought to the American colonies by ship and out to the frontier by horse, wagon and, eventually, railroad. It was used in cooking (especially desserts) and medicinal cures. The root also made an excellent ginger beer or served as a hop substitute in other beers of the time.

Ginger beer was popular worldwide, and Georgia was no exception. References to the drink exist all the way up to the early 1900s. Ginger beer is basically ginger, a sugar (molasses, cane sugar or both) and water. It could also contain a measure of grain and other herbs. Citrus, especially lemons, was a preferred ingredient when available. Ginger beer was fermented first with wild yeast and, later, domesticated yeast.

Beer on the frontier was mostly made at home, often by the women of the house. It was generally consumed on site due to difficulty of transport and spoilage. Corn beer would become an ingredient in another alcoholic beverage that could be kept longer and transported farther—namely, whiskey.

Other than household consumption of homemade alcohol, early settlers needed news from the outside world, social interaction beyond family and an occasional escape from small living quarters and hard work. They would head up the road to the nearest public house. A public house, as the term suggests, is a place where the public can gather and be served. A public house keeper was a publican. This term is also the basis for our modern

"Colonial Beer Brewing," drawing by Rose Dunning, 2013.
Courtesy of Rose Dunning.

shortened word *pub*. In the early Southeast, the public house was commonly called a tavern or inn.

EARLY ATLANTA-AREA TAVERNS

When the Georgia frontier was described by people who viewed it from atop Stone Mountain, it was portrayed as an endless forest in which a few clearings contained dwellings and farmsteads. The heavily forested area was passable on simple dirt roads used by the military, traders, travelers and new settlers. Crossroads presented an economic opportunity. Taverns sprang up along these frontier crossroads, not just in Georgia but also all across the developing nation. It was the first place in the frontier where the common folk of a place could get together to share drink, food and gossip.

The tavern would have had beds for travelers and meals and drinks for patrons. Sometimes, taverns offered entertainment. As described in *America Walks into a Bar*, most taverns were relatively uniform. They included tables where patrons could sit to drink and eat and an area where the tapster would pour drinks. Absent from these early establishments was the long wooden bar. Meals in the South would have consisted of pork (a southern staple), chicken or wild game served with corn pone, potatoes and eggs. Whiskey was common, as were varieties of wine, cider and beer. Whiskey's higher alcohol content allowed for longer storage and year-round transport. Wine was local or, if bottled well for transport, imported. Beer served in early Georgia was typically from the immediate area.

Whitehall Tavern

In 1830, the city of Atlanta did not exist, nor did most of the towns that would later compose the Greater Atlanta area. In this year, Charner Humphries and his family migrated from South Carolina to settle in the area that would become modern-day West End. Within a few years, he built the area's first tavern at the crossroads of Newnan and Sandtown Roads. It was an unusual structure because it had a coat of whitewash while most structures of this time were constructed from simple unpainted wooden planks. The tavern became known as "Whitehall" or "White Hall."

Atlanta, a City of the Modern South notes that Whitehall Tavern had the only accommodations for travelers from South Georgia to Tennessee. It served as an inn, a voting precinct for DeKalb County (before Fulton County's formation), a post office, a stagecoach stop, a mustering area for the local militia and a general community center complete with food, drink and occasional gambling. The small town of Whitehall formed around this crossroads tavern. The site functioned as the town's gathering point until Charner Humphries died in 1855. Whitehall Tavern was never used as an inn or tavern after his death and was later torn down.

The tavern was located near the modern-day intersection of Lee Street and Ralph David Abernathy Boulevard. The road leading from this tavern into Atlanta was known as "Whitehall Street." A portion of this street was later named Peachtree and became the main access through Downtown Atlanta.

Henry Irby's Tavern

The area of Atlanta commonly known as Buckhead was once Irbyville, so named for Henry Irby, who purchased 202.5 acres of land in the area and built a tavern in 1838. This combination tavern, grocery, post office and community center was located near the current intersection of West Paces Ferry Road Northeast, Roswell Road Northeast and Peachtree Road Northeast in the heart of Buckhead.

Local legend has it that either Irby or a friend shot a large antlered buck and hung the head of the buck on a post near the tavern or on the tavern itself (depending on the story). The tavern and the area became known simply as Buckhead, much to the chagrin of the town of Buckhead, Georgia, in Morgan County. Henry Irby died in 1879, and a monument honoring him was erected in the Sardis Methodist Church yard on Powers Ferry Road.

John Kile's Tavern

In 1845, the modern-day Five Points area of Atlanta was downtown Marthasville, Georgia, a mud street with a few stores and shops near the Western and Atlantic Railroad. John Kile's Tavern and Grocery was located at "Kile's Corner." Today, this site is 32 Peachtree Street Northeast in Downtown Atlanta. Little information exists about this early tavern and store. However, it was one of only two distributors of goods "both dry and wet" to Marthasville until the first saloon was introduced in 1847.

Rough and Ready Tavern

The town of Rough and Ready was a stop on the stagecoach line from Macon, Georgia, to the Atlanta area. As with many such establishments, the Rough and Ready Tavern served as the local community center, post office and a place to eat and drink. It was a cotton shipping point during the cotton boom and the gathering point for the evacuation of Atlanta during the Civil War. The tavern was also a noted hangout for Confederate spies. The tavern was torn down in 1917, and timbers from it were used in a nearby house. The town of Rough and Ready later became Mountain View, Georgia.

In 1972, the City of Mountain View (apparently still rough and ready for action) caught heat for its twenty-four-hour beer and wine sales in the otherwise dry Clayton County. Mountain View would fade, becoming a portion of modern-day Forest Park. Only a series of historical markers remains stating the history of the town of Rough and Ready and the tavern that stood near the present-day intersection of Kacoonis and Old Dixie Roads.

Pace's Tavern

Hardy Pace founded the town of Vinings (then simply know as Crossroads or Paces) around 1830, and he established the ferry across the Chattahoochee River in the area. Pace later operated a tavern primarily for drovers taking their cattle and hogs to market on the roads between Marietta and Atlanta. Paces Ferry Road is named after him and his ferry.

TERMINUS AS AN ATLANTA BEGINNING

Wagon roads became secondary to the expansion of the Southeast when the railroad took over as the preferred workhorse of change. In 1836, the Western and Atlantic Railroad line was built into Georgia. The initial portion of the railway had a terminus point in the Georgia wilderness. From all historical evidence, no one expected the terminal point to be much more than a railroad car maintenance shed.

However, a small collection of buildings, called "Terminus," soon formed around the end of the line, and this settlement became the city of Marthasville in 1843. Kile's Tavern and Grocery stood among the roughly fifteen buildings located in Marthasville in 1845. Taverns and combination tavern-and-mercantile establishments were the only common public alcohol venues in the future Atlanta area until 1847, when Mayor Moses W. Formwalt opened a building based on the latest open-room, "salon-like" format of American drinking establishment—the saloon.

CHAPTER TWO

ATLANTA'S FIRST BEER BOOM

Fermentation and civilization are inseparable.
—*John Ciardi*

ATLANTA SALOONS

In modern times, when people hear the word *saloon*, most envision a dusty western town, cowboys, gunslingers at high noon and the classic swinging doors. Although this visual has roots in truth and western lore, the saloon existed as the primary drinking establishment in the eastern United States from the early 1800s until national Prohibition went into effect in 1920. The word saloon most likely derives from the word salon. A salon is an open-area parlor or reception room. Paris, France—ever trendy—had large, popular art salons where intellectuals gathered. Early American saloons tried to capture a bit of this ambiance, using open public spaces and artwork (mostly nudes couched in mythological and historical settings). However, not all saloons were grandiose. Some were one-room shacks with dirt floors and rough lumber counters.

The common fixture of nearly all saloons was the long, lengthwise counter. This innovation was the mother of today's "bar," which now describes both the device itself and the building that houses it. Saloons of the 1800s and early 1900s typically did not have seats at the bar. There was only a long brass rail to prop the feet on and a bar to lean on. Spittoons lined the bar

Interior of Mike Shurman's Saloon, Decatur Street, Atlanta, Georgia, circa 1905. *Photo courtesy of the collection of Phyllis Levine and the Cuba Archives of the Breman Museum.*

next to the rail, and there might have been towels hanging from the bar for cleaning face and hands. The open floor or parlor areas sometimes included tables. If there were a bar with seats, it was most likely a lunch counter.

The Atlanta newspaper *Daily Intelligencer* describes the 1866 interior of the Commercial Saloon, located on Peachtree and Walton Streets in Scofield's Building (now Woodruff Park). The article describes the front room as a parlor richly carpeted and elegantly furnished with a "large and costly mirror." This room was apart from the main saloon, free from noise or disturbance and perfect for business deals. The central portion had elegant gas fixtures and a long bar with wines, liquors, cigars and strong ales. In the rear was a billiards room with four gracefully carved and ornamented rosewood billiard tables.

Although not typical of all Atlanta saloons, the Commercial Saloon had many of the standard features: mirrors; woodwork and brass fixtures; carpet, tile or hardwood flooring; gas lighting; possibly some seating; entertainment (billiards, gambling and, occasionally, "companionship"); several types of drinks; cigars; often a free lunch for patrons; and the long wooden bar. The ceilings were either pressed tin or ornately painted plaster. The buildings

An Atlanta saloon, circa 1866, after the burning of Atlanta. *Photo by G.N. Barnard, from the Library of Congress.*

themselves were usually long and narrow. Many had a front and back entrance, affording privacy if needed. Saloon owners preferred to have a corner location, which drew patrons from two angles and afforded two street entrances. Later, during prohibition periods, the corner lot would also allow views of approaching lawmen from either direction.

Atlanta: A City of the Modern South lists the first recorded Atlanta saloon as opened in 1847 by Mayor Moses W. Formwalt, a member of the Free and Rowdy political party. The opposing political party in the newly incorporated city of Atlanta was the Moral Party (no kidding). The Free and Rowdy Party was composed mainly of businessmen who owned bars, distilleries and brothels in the rowdy frontier town. The Moral Party was interested in a more temperate approach to civil construction; it relocated the seedy element of society out of the town limits, but that element didn't travel far. The outcasts settled into a nearby area called Snake Nation. Snake Nation (now Castleberry Hill) was named after the snake oil salesmen who plied their trade in the little red-light district. Snake Nation had its share of saloons, but soon these establishments would not be limited to the outcast section of town.

Decatur Street was rapidly becoming the "sporting" area of the city. Stretching east from the modern-day intersection of Pryor Street to the intersection with Hill Street, Decatur Street was early Atlanta's saloon row. Other sections of Atlanta had saloons but not at this concentration. "Black

A token from Atlanta's Railroad Depot Saloon. *Image courtesy of www.saloontokens.com.*

and white men from all over the city and the outlying countryside descended on Decatur Street on Saturday nights to visit the saloons, gaming tables, restaurants, dance halls, and brothels of 'Rusty Row,'" recounts Marni Davis in *Jews and Booze: Becoming American in the Age of Prohibition.* By 1885, the early city limits of Atlanta had over one hundred operating saloons doing $2 million in annual business. Many of these saloons were classified as whiskey saloons, but most served either whiskey and beer or beer alone, depending on their City of Atlanta liquor licenses. Through the years, the city would have many saloon regulations with most focusing on hours of operation, no sales to minors, the free lunch and no sales on Sunday.

The free lunch was a common sales technique used by 1800s saloons. Lunch items offered to paying customers typically consisted of cheese, crackers, pretzels, bologna, pickles and similar items. They were offered either on the bar counter or a nearby sideboard and were notable for their ability to inspire thirst. A technique to prompt the customer to return was the saloon token. A well-behaved paying customer might receive a coin specific to the establishment to be redeemed for the next drink or a certain amount of credit on his return.

The saloon functioned not only as a place to get a beer but also as a social gathering spot, especially for blue-collar workers. A saloon might provide mail service for those who did not have their own place, help people find a job or extend a small loan. For new immigrants, saloons run by their ethnic groups provided these services and offered English lessons and important contacts to help newcomers integrate into Atlanta.

Kenny's Saloon ad in *Sholes' Directory of the City of Atlanta, Volume II,* 1878. *Photo from the Emory University Libraries, presented in the Digital Library of Georgia.*

The modern hotel developed during the 1800s, and many of these international hotels contained saloons. The best Atlanta example was the Big Bonanza Saloon located near the Kimball House, a city block–wide hotel located at the intersection of Five Points. The Big Bonanza Saloon was also owned by a mayor of Atlanta.

Saloons were not limited to individual businesses or hotels. The first and second Atlanta Union Stations (the railroad passenger stations) had saloons. The Downtown Atlanta post office and the Gate City National Bank also housed saloons. The Magnolia Restaurant was operating a saloon in Downtown Atlanta as early as 1854.

These businesses were often named after the owner, a trade group or a famous European drinking establishment. The Globe, House of Lords, Shamrock Saloon, Le Bon Ton Saloon, Alhambra, Mechanics Saloon, Planters Saloon, O.L. Pease's Saloon, C.J. Vaughan's, Campbell's Saloon, Kenny's Saloon and Ponce de Leon Saloon were a few of the hundreds of saloons in nineteenth-century Atlanta. The Girl of the Period, a saloon located at Number 16 Marietta Street, played a role in young Doc Holliday's life before his western exploits with Wyatt Earp and the gunfight at the O.K. Corral.

EARLY ATLANTA BREWERIES

America's drinking establishments were growing and changing, as was American beer. The 1840s to 1880s saw an influx of German immigrants, and with them came the brewing technology to produce lager beer. Specialized railroad cars cooled by blocks of natural ice came into use, allowing perishable goods to travel farther. The increase of saloons and the influx of the German beer culture stimulated beer production in America. Atlanta was no exception. German immigrants opened saloons, beer gardens and retail shops and formed social societies, such as the Turnverein. The Turners were a German athletic society known for its singing and beer drinking. Concerts and other social events were held with beer and favorite foods reminiscent of the homeland.

If one man could be named the godfather of brewing in Atlanta, it would be Egidius "Edgion" Fechter. He was born in Baden-Wurttemberg,

The grave of Egidius and Dionis Fechter in Atlanta's historic Oakland Cemetery. *Ron Smith.*

Germany, in 1824 and came to Atlanta with brother Dionis in the early 1850s. Egidius started the area's first brewery near Howell Park in West End (which would later be incorporated into Atlanta). Unfortunately, the name and exact location of his brewery is lost to history. Egidius Fechter's Howell Park brewery could have introduced lager beer in the Atlanta area. This early lager was most likely *schenk*, or "steam beer," which could be produced in three days without refrigeration but did not have the shelf life of a fully lagered (clarified, chilled and cellared) beer, according to the technical papers of German brewers in the 1800s. (An 1878 reference in the *Macon Telegraph and Messenger* could be one of the earliest Southeast mentions of this beer style.)

By 1858, Dionis Fechter and a Mr. O. Kontz (O. Kontz & Company) were operating the City Brewery at the intersection of Marietta and Pine Streets, near the Western and Atlantic Railroad line (roughly current-day 556 Marietta Street). They also manufactured lager beer, most likely the schenk style, since no mention of cellars for storing the beer was found connected to this location.

Both breweries operated at least three years until the outbreak of the Civil War in 1861. Military records and historical timing suggest that Egidius Fechter left his Howell Park brewery behind due to the war. There is no evidence of his operating the Howell Park brewery after the Civil War. The City Brewery appears to have stopped production for a few years during the time of the Battle of Atlanta and Sherman's occupation, though both breweries appear to have survived the burning of Atlanta.

An August 1865 newspaper ad run in the *Daily Intelligencer* (mere months after the end of the Civil War) states that the City Brewery, well known in past history, was ready to resume business at its old location on Marietta Street and lists "O. Kontz & Fechter" as the continued owners of the brewery. However, Kontz would not stay in the business much longer. By the following month, Dionis Fechter, Michael Kries and F.A. DeGeorgis owned the City Brewery. The brewery would continue to operate for a short time at its Marietta Street location while a new brewery was built between 1866 and 1868 at the corner of Courtland and Harris Streets (location of the modern-day downtown Hilton Hotel). By 1868, Edward Mercer had purchased an interest in the brewery, and Fechter and Mercer's City Brewery was producing award-winning lager beer and possibly porter ale.

The City Brewery (detailed later), Atlanta's highest-production brewery, was not without competition. The 1860s would be a boom decade for Atlanta breweries.

Fulton Brewery

An article in the *Daily Intelligencer* indicated that the "Fulton Brewery is completed and the Fulton Brewery Saloon will open on the Fourth of July 1866." Adding more details about the brewery's offerings, the article states that "the brewery produces lager beer and ale...the saloon will have lager for 5¢ per glass and lunch on call." The old Fulton Brewery was located somewhere between the intersection of current-day Glenn Street and Humphries Street and the intersection of Ralph David Abernathy Boulevard and Metropolitan Parkway. It was operated by Michael Kries (formerly of the City Brewery in 1865). Mr. G. Orthey appears to have operated the Fulton Lager Beer Saloon, located in the Scofield Building (now Woodruff Park). Articles in the *Daily Intelligencer* from July to November 1866 state that the Fulton Brewery made beer "by [its] own manufactory" and the brewery and saloon would "take any quantity orders and they will supply families in small quantities by keg, gallon, or otherwise."

This reference to small orders is one of the first indications of an early growler business in Atlanta. Family members would show up at a retailer or brewery with a container and get beer to take back to their residences. Also, businesses could send a runner with a pail to get beer for the workers. These containers came in many sizes and were made of wood, metal, stoneware or other materials. In *The Old-Time Saloon* by George Ade, he mentions "customers who 'rushed the growler' and came in to get their malt products in pitchers and buckets." There are many urban legends as to the origin of the term growler. None of them has been verified; however, all the myths seem to center on carbon dioxide pressure and the beer bubbling and growling. The City of Atlanta had regulations regarding this type of container as early as 1899, stating that the beer could not be consumed from the growler while at the saloon.

The Fulton Lager Beer Saloon appears in the first written references to an Atlanta saloon being a tied house. A tied house was an establishment that sold only a single brewery's beer. In exchange for built-in distribution for the brewery, the establishment usually gained financial support from the brewery to offset the high cost of county and city liquor licenses. The arrangement was mutually beneficial as long as the product sold. Later newspaper references indicate that the Atlanta City Brewery had many tied houses over the years. Competition could be one of the reasons that the Fulton Brewery was out of business by the early 1870s.

Atlanta Steam Brewery

Newspaper ads for Spencer and Company's Atlanta Steam Brewery, which made "ale, porter, and beer," begin to appear in the *Atlanta Constitution* in 1870. The brewery is also listed in that year's *Atlanta City Directory*. The entry states the business was owned by C.A. Goodyear and claims it to be the "only ale brewery south of Richmond, VA." The *Weekly Atlanta Intelligencer* published an 1871 interview with the proprietor. Goodyear mentions the brewery's production of ale and porter and that the problem of the summer heat was not a concern to the Atlanta Steam Brewery. The interview mentions that the brewery's beer barrels were made from local trees cut down from within a mile of the Kimball House in Downtown Atlanta. The owner was having barley planted locally, and his agents (salesmen and distributors) were known from "sea to gulf and in Alabama and Middle Tennessee." The brewery's beer was being sold at hotels, stores and boardinghouses. Interestingly, the newspaper article clearly reflects Mr. Goodyear's preference to have no northern business partnerships.

The term *steam* again appears associated with beer production, this time in the name of the brewery. Although breweries of this age were operated by coal-fired steam engines, it is unlikely that the name Atlanta Steam Brewery was applied for this reason since this method of operation was standard for larger-scale producers and would not be a differentiating quality. A better hypothesis is that the term was connected to the likely use of a cool ship—a large, shallow and open-top basin used to rapidly cool the boiling wort (prefermented malt liquid)—which produced a cloud of steam rising off the cooling liquid. Further, if the brewery promptly packaged and distributed lager beer from the cool ship, then it was producing schenk or steam beer. The alternative hypothesis, then, is the Atlanta Steam Brewery was named for both the method of cooling and for the style of beer produced.

The 1874 *Atlanta City Directory* contains the last mention of the Atlanta Steam Brewery. The facility was likely located somewhere near 503–529 Glenn Street. This location is very close to the location of the Fulton Brewery. The operating timeline and property records indicate that these were separate breweries (independently owned and operated).

Georgia Spring Brewery

Local newspaper references to the Georgia Spring Brewery first appear in 1867. The brewery, owned by Frederick Richter and a Mr. Orthey (most likely G. Orthey of the Fulton Lager Beer Saloon), operated from around 1867 to 1874 or 1875. The book *Atlanta As It Is* names the brewery and mentions that it made an "immense amount of lager beer." A story in the July 8, 1873 *Atlanta Daily Herald* tells of a reporter's trip to West End to the brewery and a nearby park on the Atlanta streetcar line. He describes the brewery as having a dancing platform, a saloon, beer tables, a shooting gallery, a flying Jenny (Ferris wheel) and a "small lake and boats." On the other side of the lake was the "Richter Mansion."

Several 1867 newspaper references to the Georgia Spring Brewery and the 1880 obituary of Richter mention that he "built up and improved the old brewery." This old brewery is variously described as being in West End, near West End, near Camp's Mineral Spring or in Howell Park. The location of the original brewery, which was improved to make the new Georgia Spring Brewery, is consistent with the location of Egidius Fechter's Howell Park brewery, Atlanta's first recorded brewery. Little evidence remains to pinpoint the exact location or the likely transition of the brewery from Mr. Fechter to Mr. Richter shortly after the Civil War.

WHOLESALERS AND DEPOTS (DISTRIBUTORS)

Not all early Atlanta beer was produced locally. Between 1859 and 1862, the Etowah Brewery in Etowah, Georgia, was selling beer in Atlanta through its "agent" Hans Muhlenbrink. He was a noted saloon owner, liquor wholesaler, fireman of Atlanta Fire House Number 1, member of the Atlanta German-American community and charter member of the Atlanta Turnverein. Muhlenbrink is also credited with bringing the first billiard table to Atlanta in 1857.

Another famous saloon owner and liquor wholesaler was Michael E. Kenny, an Irish immigrant and businessman. He owned and operated the Chicago Ale Depot on Pryor Street. The Depot, as the name indicates, received beer from Chicago-area distributors. As with most depots, the business offered beer, wine, whiskey, other spirits and cigars (sometimes spelled "segars"). Mr. Kenny's fame would be established by his unfortunate

Statue of Michael E. Kenny in Kenny's Alley, Underground Atlanta. *Ron Smith.*

death due to a jousting accident during the 1870 Georgia State Fair. Kenny's Alley running east from Pryor Street to Central Avenue, at the heart of the Atlanta underground, is named after him.

The Railroad Ale House, a saloon and wholesaler, was a major importer of beer into Atlanta starting in 1870. Its ads in the *Atlanta Daily Intelligencer* mention that it supplied draught ale (wooden kegs) from Philadelphia and imported porter

Workers at a bottling works in Atlanta, circa late 1800s. *Photo courtesy of the Kenan Research Center at the Atlanta History Center.*

and ale from abroad. This alehouse also carried a supply of cider and cigars and offered the patron-enticing free lunch. In mid-1870, the business received a large refrigerator from Cincinnati, making it one of the first small businesses to have mechanical refrigeration. The Railroad Ale House was located at 54 East Alabama Street, an address that is also now part of Underground Atlanta.

Other beer retailers, such as the Cincinnati Beer and Ale Depot, Kenny's Ale Depot and McGuire's Ale Vault, received beer from other areas of the nation: Ohio (mainly Cincinnati's Over-the-Rhine area), Milwaukee, Indiana, Virginia and Chicago. Imported beer from Great Britain, Germany and other nations also arrived in Atlanta as refrigeration, shipping and railway lines improved.

Local bottling companies, such as J.H. Spilman, Jones and Company, and other minor concerns, such as a small beer-bottling business at 182 Decatur Street, received beer from local, national and international sources in draught via train or local wagon and bottled the beer for Atlantans to consume. Whether the bottled beer was labeled with the brewer's name or the bottler's name is often lost to the mists of history. The Atlanta City Brewery exported a sizable volume of beer as well. The majority of this exported beer was delivered to southeastern states and possibly a few other countries. This exported beer consisted mainly of lager beer in stoneware bottles tightly packed for shipping.

Atlanta Beer Styles, Methodology
and Terminology (1850–1916)

Save for use of the words ale and lager to describe the yeast used, resulting fermentation and storage method, most modern categorization of beer styles did not occur until the 1970s. The terms *ale* and *lager* were coined before significant research was conducted on yeast and are based on the visible signs of fermentation. Ales are typically brewed with top-fermenting yeast strains (i.e., the top of the fermenting beer has a large bubble layer called *krausen*). Ale yeasts usually live at warmer temperatures and produce beer at a rapid pace. Lager—a term taken from the German word meaning "to store"—typically has a longer and cooler fermentation period and is brewed using bottom-fermenting lager yeast with a less persistent krausen. As with all things biological, there are gray areas, and some beers mix elements of these methods.

Beer advertisements and brewers' descriptions from the late 1800s and early twentieth century are based solely on the individual's knowledge and associated regional history. The term *ale* can be used to describe various types of beer made using ale yeast. The term *beer* is most often used to describe a lager beer, though its use is not restricted to lager. In other words, if an ad lists "Porter, Ale, and Beer," it most likely means a porter- or stout-style ale, a pale or similar ale and a lager. Even the term *liquor* was used in a broad sense. It typically meant distilled alcohol, but this was not always so. Liquor was also used to describe all alcoholic beverages including beer, wine and cider.

While various ales, including porter, were made in Atlanta during the 1800s, the influx of German immigrants started a lager revolution. This is consistent with United States history as a whole. The wholesalers and depots introduced other German-style beer into Atlanta. One of the largest beer producers in the late 1800s, the Christian Moerlein Brewing Company of Cincinnati, shipped beer across the United States and exported to other countries. Ads for its beer appear in early Atlanta newspapers, naming the local retailers that sold its products. At one point, the Atlanta City Brewery found itself in a lager price war with Christian Moerlein and Pabst Brewing Company.

The Crescent Brewing Company of Aurora, Indiana, sold "Vienna Beer" in Atlanta during this time. It was most likely an amber malty lager based on Austrian brewing techniques (Vienna lager). Occasionally, bock beer makes an appearance in Atlanta beer history. Bock is a higher alcohol by volume (percent ABV) lager beer that was made seasonally and originated in the

German city of Einbeck. However, in most cases, the older texts do not indicate what style of bock beer is being offered, and style must be inferred from the timing of availability. In one case, a May 1875 *Macon Telegraph and Messenger* article declares the city will have bock beer from the Atlanta Brewery available soon. It goes on to describe the brew as clear and bright with a splendid body like the finest liquor and to say that it is only available in the month of May. Hence, the beer was likely a maibock.

Even though lager was the rage in the 1800s, ales were still brewed, and records show that they were exported from and imported to Atlanta. Most local and national brewers were offering some form of ale, which was often a porter. A December 1865 *Daily Intelligencer* ad for F. Corra and Company Retailers mentions a beer style rarely seen in early Atlanta—India pale ale.

Beer in the 1800s and early 1900s was bottled or barreled. Metal kegs did not exist until much later. Beer barrels came in a variety of sizes. A railroad manifest in the March 26, 1887 *Atlanta Constitution* lists full barrels, half barrels, quarter barrels and eighth barrels. Beer barrels were made with extra-thick staves to resist bursting from the pressure of naturally occurring carbonation. They were typically made from local oak trees and produced by a cooper working for the brewery or at an independent barrel works.

A barrel could weigh as much as 100 pounds empty and 350 pounds when full of beer. These barrels were delivered on flatbed wagons called drays. Accidents with shifting barrels were not uncommon, and crushed hands were often the result. Occasionally, a barrel exploded from internal pressure. A story in the *Knoxville Press and Herald* recounted a man suffering from a broken jaw, missing teeth and gashes to his face from an exploding ale barrel. The force of the explosion also drove shards of wood into the plaster ceiling of the saloon.

Early beer bottles came in various styles and sizes before the standardization of the crown-top bottle beginning in the late nineteenth century. Early ginger beer was bottled in stoneware. Ginger beer was so popular that this style of stoneware bottle is called a "ginger beer bottle." Export beer bottles were typically larger and made from either glass or stoneware. Many 1800s glass beer bottles, especially early lager bottles, were modeled after wine and champagne bottles. These were often corked and caged, the cage keeping the cork and pressurized contents in place during transportation and storage.

Hutchinson-style bottles were popular with soda makers and were also used by breweries into the early twentieth century. Early screw-top bottles and variations of the lightning stopper (swing-top) bottle were also used. In addition to the industry standard crown-top bottle, today's breweries still rely on screw-top, swing-top and 750-milliliter corked-and-caged bottles.

Although a variety of beer barware existed in late 1800s Atlanta, the container most referenced in print is the schooner. Advertisements from the era show an assortment of stemware, steins and mugs made of ceramics, metals and glass. During the early part of the twentieth century, saloon beer was often dispensed in eight-ounce pours in highball or tumbler-style glasses. A snit was a small taste of beer (up to three ounces) served in a small glass, similar to the modern sample in a beer bar. Custom metal serving trays, company-embossed corkscrews and other stylized barware and breweriana from this time still exist as branded collectibles.

Beer was not always imbibed cold, or even cool. It was preferred cooler in the summer months, but ice was hard to come by, especially before the days of mechanical refrigeration. It could be cellared to keep it cooler, but beer was often served at room temperature. During chilly seasons, ale was mulled much like wine. Tin and copper ale warmers, or any available vessel, were used to warm the beer. Winter spices were added to the hot or warm ale.

Whether warm or cool, ale and ginger beer were used in compound drinks. An example of this approach was the Shandy Gaff, made with a fifty-fifty mix of ginger beer and ale. The Velvet was half champagne and half porter. An 'Arf and 'Arf (Half and Half) was an equal mix of porter and ale, and the humorously named Mother-in-Law was a mix of old ale and bitter ale. An old preparation that now seems rather odd was the practice of floating a piece of toasted or sweetened bread on top of some of the compound beer drinks.

By the 1880s, Atlanta's residents were consuming a variety of beer-based drinks in saloons, taverns, beer gardens and hotel bars and were also taking beer home in growlers. However, not everyone was happy about the availability of alcohol in any form or quantity. A temperance movement began to persuade Americans to voluntarily abstain from drinking alcohol-containing beverages. Soon, the movement would pursue a different tactic: legal mandate.

CHAPTER THREE

PROHIBITION COMES TO ATLANTA

In gentle Georgia, my dear old state,
There will be no whiskey in 1908.
There'll be no whiskey, there'll be no beer
In dear old Georgia for many a year.
—Temperance ditty from 1908
Remembered by Irwin Shields in
Living Atlanta: An Oral History of the City, 1914–1948

The story of Prohibition in Atlanta is a complicated tale. We see today a confusing patchwork of alcohol blue laws (laws to enforce religious standards), relics of a moral and social war that raged across both Georgia and the nation. National Prohibition is somewhat familiar to most, but the individual state-legislated acts of alcohol prohibition are lesser known and often more complex. Georgia's prohibition period—similar to America's as a whole—divided people among those advocating prohibition of alcohol, those against prohibition of alcohol and the moderate majority. Journalists of the age commonly called anyone associated with temperance and prohibition *Prohibs* or *Drys*. The anti-prohibitionists (those who wanted to protect access to alcohol) were often called *Antis* or *Wets*.

The term *Prohibition* conjures images of flappers, speakeasies, cocktails and gangster violence, typically centered in northeastern cities like Chicago and New York. Prohibition in Atlanta did not entertain with socialite scandals or infamous nightclub speakeasies. Modern Americans have a romantic view of

Georgia Woman's Christian Temperance Union (WCTU) parade float. *Photo courtesy of Georgia Archives, Vanishing Georgia Collection, image number dec014.*

the lawlessness and social rebellion that made up Prohibition. In truth, state and federal prohibitions would hold up an unkind mirror to Atlanta's social problems. It would pit urban areas against rural areas, widen an already existing racial chasm, reignite anti-immigrant sentiment, spotlight religious inequalities and further complicate the interaction of social classes.

Atlanta's dry movement was driven by temperance groups consisting mainly of Protestant evangelicals. The Georgia chapter of the Woman's Christian Temperance Union (WCTU) was formed in 1883, riding the latest wave of America's Protestant "awakening." Temperance was originally defined as moderation or self-restraint in eating or drinking, in this case, moderate drinking of alcohol. Temperance began as a movement to rid the church of drunkenness but became a national wave to help relieve all members of society from the dominance of alcohol in their lives. In the final decades of the 1800s, the trend of moral persuasion and moderation among the temperance leaders gave way to teetotalism and legal influence.

The Anti-Saloon League (ASL) was founded in Ohio in 1893. Its influence spread rapidly, and by 1895, it was a national organization called the Anti-

Saloon League of America. The ASL originated pressure politics in the United States. The organization would launch an impressive print campaign, and its ability to slander any politician who even remotely supported alcohol trade was legendary. The de facto leader of the Anti-Saloon League of America, Wayne B. Wheeler, publicly admitted that the ASL would support a politician who drank alcohol, as long as he voted against other people's drinking of it.

The ASL's Ohio-based printing presses turned out tons of printed materials (posters, pamphlets, pledge vouchers, magazines and other paraphernalia) each year. The group printed anti-saloon, anti-brewery and anti-alcohol ads and articles in local newspapers as frequently as possible.

Beyond pressure politics and sheer volume of printed material, the effectiveness of the ASL resided in its modern corporate-like structure and razor-sharp focus. The ASL understood that most alcohol sales were made at the saloon. Therefore, it chose the saloon as the target of its combined might. The Georgia branch of the ASL was formed in 1905 and followed the lead of the Anti-Saloon League of America. The national organization was not above dominating other temperance groups such as the WCTU and the Prohibition Party to gain its ultimate goal—complete eradication of alcohol consumption.

The WCTU and the ASL were not the only temperance organizations in the South, just the largest and most influential. Other groups, including the Sons of Temperance, Cadets of Temperance, Cold Water Army, Daughters of Temperance, Knights of Jericho, the Good Templars, the Rechabites and the Templars of Honor, had presence in the southern states.

The anti-Prohibitionists were a diverse group and not as well organized as the Drys. Very few formal organizations existed among Atlanta's Wets. Hans Muhlenbrink, noted saloon owner and liquor dealer in Atlanta, organized the Atlanta Liquor Dealers' Association in 1870. (Whether he formed the association in response to the rising tide of temperance sentiment in Atlanta is unknown.) The Atlanta City Brewery was a member of the United States Brewers' Association, but the association was slow to respond to the temperance movement. In 1885, the anti-Prohibitionists, along with the Mutual Aid Brotherhood, the political arm of the Knights of Labor, formed a "focus group" and, later, a political ticket against Prohibition.

The Model License League was formed in 1908 as an alternative to temperance. The league promised to reform the liquor licensing system as a means to control trade in alcohol without prohibition and offered a surprisingly modern view of regulation. Unfortunately, the Drys were set

Temperance crowd gathered for a vote on local prohibition. *Photo courtesy of Georgia Archives, Vanishing Georgia Collection, image number low104.*

on their narrow view of alcohol as the root cause of society's woes and dismissed the Model License League as promoting evil. However, the issue wasn't entirely a moral issue—it also presented monetary, economic and social aspects.

The largest alcohol retailers in Atlanta during Georgia's prohibition era were the Atlanta City Brewery and the Kimball House Hotel (including the saloon within, the Big Bonanza, owned by Atlanta's mayor). The alcohol industry in late 1800s Atlanta encompassed hundreds of saloon owners, retailers, depots, warehouses and associated businesses. Supporting trades included coopers, bottlers, refrigeration and ice vendors and coal distributors (for fuel). These business owners, their workforces and many politicians worried that if the alcohol industry and its sizable contributions to Atlanta were to disappear, the city would suffer.

During Prohibition, news journalists and religious leaders became not only increasingly racist but also anti-immigrant. Building on the foundation laid by the Nativist Party in the 1850s, they portrayed African American Wets and immigrant saloon owners who served them as foreigners who preyed on America. Prohibition had become a battle over not only alcohol but also more generally about who was a "real American."

The majority of Atlanta's Jewish and Catholic populations saw no moral problem with the consumption of alcohol. At the turn of the century, a stretch of Decatur Street was early Atlanta's international district, consisting of Chinese, Greek, Italian, Syrian and European Jewish businesses and residences. Often the owners of the "Colored Saloons" and the very few integrated saloons on Decatur Street were Jewish businessmen and women, many of Russian descent.

Early Atlanta's European immigrants valued personal liberties based on a different cultural set than that of the Drys. Alcohol consumption was simply part of the social structure. Historically, in many European societies, men and women drank together over meals in private and in public. Immigrants continued these social traditions when they arrived in the United States. This public intermingling of the sexes over alcohol was seen as amoral by the Protestant evangelical Drys and widened the rift between Drys and the new immigrant groups.

THE ROAD TO STATE PROHIBITION

Temperance and prohibition of alcohol in the Atlanta area began to take shape before Georgia was a state. The early Georgia Colony Charter of 1732 barred liquor dealers, lawyers and Catholics from the colony. Several attempts were made to organize temperance movements in Georgia, but it was not until the 1850s that real organization was gained. Had the Civil War not occurred to interrupt the effort, the temperance movement in Georgia would have had a far earlier start.

The WCTU's and ASL's arguments were not only biblically and emotionally based. These groups also pointed to the rising consumption of alcohol in America for its effect on the family both in domestic violence and economic ruin. Indeed, alcohol consumption was on the rise, especially in the South. The Civil War had crippled the Southern economy, and the years following posed a physical and emotional depression to many.

The temptation to drink was exacerbated by new industrial technology that allowed more output of any manufactured product, including alcohol. The Atlanta City Brewery faced increasing competition during this period, as beer from large-capacity or consolidated out-of-state breweries flowed into the city. Pabst Brewing Company in Milwaukee owned a tied house in Downtown Atlanta at the turn of the century, and in 1895, it planned on building a large brewery in Atlanta.

The city saloon, though a valuable blue-collar social institution, was not family friendly. During much of Atlanta's history, saloons were segregated by both race and gender. Atlanta saloons, as with saloons across the country, were a homosocial public space. Only one gender entered and socialized at the saloon: men. Even if a gentleman were married, the saloon might be the cornerstone of his social life. If a woman frequented a saloon, she appeared to be a "low woman" (a prostitute) in the eyes of southern Protestants. The women frequently seen near a saloon were wives or daughters picking up a growler of beer for the household or liquor for cooking. Such transactions were commonly completed at the side door.

Most people recognized there was a growing national problem. The issue lay with how the problem was to be solved. An initial step was clearly defining the matter—whether it was an alcohol legality problem, a moral issue, a social crisis or something more complex. The Drys were convinced it was a "saloon problem," and they were going to fix it.

One of the temperance movement's legal alcohol-control techniques was to vote for higher costs on liquor licenses. The practice of alcohol licensing in Atlanta is nearly as old as the city itself. Early temperance advocates voted to increase the fees to operate saloons, hotel bars and other outlets in hopes of closing or limiting the number of establishments selling alcohol. By 1887, an annual liquor license in Atlanta cost $1,500, and a beer license cost $100. These costs would be astronomical in today's dollars.

By the late 1800s, temperance advocates deemed increased license costs a failure. The number of alcohol vendors did not decrease; businessmen simply paid higher fees and continued to operate. To the chagrin of the Drys, the city coffers grew due to the increased fees. This allowed the Wets to argue that the city's sound infrastructure and services (including schools, police forces, fire department and so on) were due to licensing fees and other taxes paid by their industry.

Another Dry regulation technique was to couch the alcohol problem as a local community problem. Through the early years of the 1880s, Georgia Drys urged lawmakers in each county to exercise the right to impose prohibition of alcohol—a concept called local option. A statewide local-option law was pushed through by the original proponents of higher licenses: the WCTU and vocal church leaders. By 1884, many Georgia counties were partially dry (had some limitations on alcohol sales). The temperance movement relied heavily on the rural vote. As a result, the movement tended to glorify farm life and demonize the industrial cities.

In 1885, a campaign was started to vote for prohibition in Fulton County and therefore the City of Atlanta. Both the Wets and Drys held mass meetings in Atlanta to promote their platforms. Pamphlets were handed out expounding either the virtues of abstinence or the economic pitfalls of cutting off a viable tax source for Atlanta. This political rhetoric, later common within prohibition efforts across the state, began in Atlanta. The anti-prohibition pamphlet titled *An Appeal to the Voters of Atlanta* stated, "They even oppose baseball, theaters, card playing, dancing, and what not." The Drys circulated a ticket with an angel on it, and their platform became known as the Angel Ticket.

Exercising the local-option law, Fulton County held its first prohibition vote in November 1885. With strong rural county support, the prohibition measure passed by a slim margin, even though most Atlanta precincts voted against prohibition. The law banned the sale of whiskey and ardent spirits in any quantity under a quart. This prevented the sale of single shots of whiskey, the most common saloon drink. It allowed for sale of only "native wines," thus aiding Georgia wine farmers (the rural vote) by banning imported wines.

The wild card within this new law was beer and the question of the Atlanta City Brewery. The prohibitionists generally agreed the brewery could brew all the beer it wanted but could not sell it in Fulton County. After much debate and further ordinances, the city council ruled against local sales by the brewery. This decision was made despite the argument by the brewery that beer sold outside the county was legal to bring into the city. Representatives of the Atlanta City Brewery also argued that existing beer was the property of its national and international investors and the City of Atlanta had no legal authority to confiscate the investors' property.

Every Atlanta newspaper was soon carrying liquor ads from every point outside Atlanta. Alcohol could be bought by telegraph or mail correspondence from other wet Georgia counties (or other states) and started streaming into Atlanta via railroad. So-called jug trains became commonplace. Everyone else was getting rich off Atlanta's self-imposed prohibition. The Atlanta City Brewery made its profits by distributing much of its beer to nearby towns (notably Macon), where the beer was in turn sold locally and transported back into Atlanta by rail or horse-drawn wagon.

Many of Atlanta's major retailers, existing private clubs and other businesses that could afford legal aid found loopholes and survived; many others did not. A large number of saloons, restaurants and other venues simply became wine rooms, selling their own interpretation of allowable native wines. Atlanta's legal system had to deal with outright defiance of the new law, determining whether the new wine rooms were selling wine that was legal.

There was also an onslaught of prohibition-era beverages labeled "Non-Alcoholic" and "Non-Intoxicating." Noteworthy 1886–1888 Atlanta prohibition beers were New Era Beer and Rice Beer. Exactly what these two beverages were cannot be determined. Prohibitionists contended they were lager beer labeled under code names. It is likely that these drinks were actual beer or a beer-like base that could be spiked with a hidden stash of pure alcohol to bump up the alcohol content of each drink. Supposedly, patrons could use a system of code words and gestures to indicate they wanted the "real thing" (full lager beer) or for the new product to be "enhanced."

In times of alcohol prohibition, the term *speakeasy* was common in the Northeast. The terms *blind tiger* and *blind pig* were used more frequently in Atlanta. A blind tiger was often a person's residence. The terms come from a ploy to get around the restrictions on selling alcohol. A person could charge a group of people to view an animal oddity (the blind pig) and throw in a drink for free. The term quickly evolved into any illegal selling of alcohol by a (typically) small-quantity operation. The term could also apply to the sellers themselves, who were called tigers.

A roving tiger or moving tiger arose when several houses (or any location) would have a block party. The alcohol would move throughout the evening, and the crowd would mingle between the houses, occasionally visiting the one currently holding the booze. A pocket tiger was a person who carried a flask in a pocket and gave out swigs on the move. This term is comparable to bootlegger, which refers to alcohol being hidden in the boot.

With Fulton County going dry in 1886, blind tiger became a familiar term not only with law enforcement but also more broadly in newspapers

A drunkard dancing with a blind tiger and losing his money. *Atlanta Constitution*, September 28, 1909. *Image courtesy of www.Fold3.com.*

and with the public. Blind tiger arrests for illegal sales of liquor and beer numbered in the hundreds, ranging from a suspected beer blind tiger at a Broad Street residence to Ms. Lucy McCall, an entrepreneurial sixteen-year-old girl selling liquor wholesale.

The year 1887 held a hotly contested second vote on Fulton County prohibition. Atlanta was divided, and the conflict was visible. The city's Drys walked the streets sporting blue ribbons, a sign of their voting intent. The Wets, in turn, had selected a red ribbon for their cause. Family members were in disagreement, and street fights over the issue were not uncommon. Finally, after the city waited with bated breath for the results, on November 26 the citizens of the city of Atlanta and Fulton County voted for the county to return to wet status. The local politicians, seeing how much damage the extremely partisan issue was doing to the city, agreed not to pursue a two-year option on the original law.

Atlanta's Drys were not going to accept defeat under any circumstances. Temperance sentiment was still running high in the city as well as the rest of the United States. By 1901, temperance doctrine was widespread. Every state required that its schools teach "Temperance Instruction" as part of the curriculum.

Through the turn of the century, more and more Georgia counties went dry. Temperance leaders across America were focused beyond the local county level, setting their sights on prohibition at state and national levels. State prohibition would eventually force Atlanta to revisit dry status and provide a steppingstone toward what many thought impossible—national Prohibition, the holy grail sought by the temperance movement.

The 1906 Atlanta Race Riot would provide just the opportunity the Prohibitionists were looking for. The bloody riot gained not only national but also international attention. The world was watching to see the outcome of this tragedy. The Drys blamed Atlanta's saloons for the outbreak of violence, specifically the establishments along Decatur Street's Rusty Row, which included the integrated African American and immigrant-owned saloons. This opinion was shared by city politicians—despite the fact that the mob perpetrating the violence coalesced from the white saloons. The stage was set to do something about Georgia's saloon problem. Riding the racially driven anti-alcohol fervor that had dominated the Atlanta press throughout 1906, lawmakers drafted legislation to enact statewide prohibition.

In 1907, the Hardman-Covington-Neel Bill made its way through state legislatures. This bill would lead to mandatory statewide prohibition in 1908, with Georgia becoming one of the first southern states to pass such a law. An article in the July 24 *Atlanta Constitution* sarcastically observed, "On the eve of passing the Hardman-Covington-Neel Prohibition Bill, the majority of the [Georgia] house of representatives yesterday afternoon consumed seven kegs of beer."

On the stroke of midnight as 1908 was heralded, Georgia became a dry state. Prohibitionists in Atlanta threw huge New Year's Eve parties (dry, of course), and church bells across the city rang in victory. A crowd gathered at the Baptist Tabernacle, and the gold pen that was used to sign the state prohibition bill was displayed.

NEAR BEER AND OTHER ALTERNATIVES

As before, dry was a relative term. While the *1908 World Almanac and Encyclopedia* declared, "Georgia becomes a prohibition State on January 1, 1908, and the law is so drastic that wine cannot be used at communion services in churches," the law was not all encompassing. Allowances and loopholes in the new law included near beer, pharmacy-prescribed alcohol and locker clubs. Near beer was defined as a nonintoxicating beer. This did not mean the beer had no alcohol, it was just assumed to contain a nonintoxicating amount. The new legislation banned any drink that "if drunk to excess, will produce intoxication."

Saloons quickly reinvented themselves as near beer saloons. In the same year, a City of Atlanta ordinance made it unlawful for any person to sell near beers or any other drinks intended to be used as substitutes for "lager beer" without a license (at a cost of $200). Merchants could not sell near beer on Sunday, and sales to minors or alongside a free lunch were forbidden. Near beer kegs and bottles had to be stamped with the manufacturer's name. The Atlanta Brewing and Ice Company (formerly Atlanta City Brewery) adapted to the new state of events and sold its near beer to around 120 venues in the city.

Locker clubs were private clubs where the alcohol was kept under lock and key (a locker). As drinks were only served to club members, locker clubs skirted the law by not selling to the public. The legal gray area arose from guests and temporary members also partaking of the club's alcohol. During the period from 1900 to 1910, at least twenty well-known locker clubs operated in and around the city of Atlanta. The Drys would fight for seven years attempting to shut them down. However, the Drys met with firm resistance from lawmakers, most likely due to the prevalence of their membership in these private clubs.

Atlanta's Drys were growing increasingly frustrated with the inability to rid the city of alcohol. A July 1908 *Atlanta Constitution* article exemplified such frustration:

> *Saloons on such streets as Decatur, Peters and Marietta, which are selling near-beer, present even a much more objectionable appearance than did the old liquor saloons when they were thronged and packed with men on certain nights in the week. In many of the near-beer saloons both men and women congregate, sitting around tables, drinking the near stuff with all the riotous enjoyment and hilarious feeling that would be expected if the straight beer was being used to quench thirst.*

This sentiment is, most likely, a reflection of the Drys' reactions to the immigrant and personal liberties crowd that made up early Atlanta's international district.

In addition to their objections to both sexes drinking together in public, the Drys were obviously skeptical that near beer was not intoxicating. By modern standards, they would be correct. In June 1910, Atlanta adopted an amendment that near beer could not exceed 4 percent ABV. Ironically, the majority of beer currently produced in America by macrobreweries is within 1 percent ABV of 1910 Atlanta near beer.

At the time the near beer amendment was being adopted, an issue of the *Atlanta Constitution* declared in a front-page title, "Fight Is On against the Sale of Near Beer." Atlanta's Drys, led by the ASL, rallied to fight against their new focus, the near beer saloon. The *Atlanta Constitution* was loaded with lists of saloons that were submitting a public notice of application for this type of license. Atlanta in mid-1910 had around 201 active licenses for the sale of near beer.

GEORGIA TIGHTENS THE LID ON ALCOHOL

It wasn't enough that the vast majority of alcohol was already illegal in Georgia, the Drys wanted it absolute—capital *T* Teetotal. A November 1915 *New York Times* article quoted a radical Georgia Dry as saying he wished "the smell of liquor illegal in Georgia." In 1916 those who held this sentiment were one step closer to getting their wish.

Effective May 1, 1916, the latest amendments to Georgia state prohibition banned locker clubs, abolished breweries and near beer saloons and tightened the policing of blind tigers. The anti-shipping portions of the law prohibited delivery of liquors to any place of business, including restaurants, hotels and clubs. Citizens of Georgia could receive at their private residences two quarts of spirituous liquor, one gallon of wine or six gallons (forty-eight pints) of malt liquor every thirty days. The alcoholic drinks had to be shipped from outside the state and consumed within the thirty days or destroyed; no storage was allowed above the specified volumes. The deliverer of the shipment had to gain a sworn statement from the receiver that it was for private use only and that the resident was not a drunkard.

The saloon as an Atlanta drinking establishment ceased to exist. Eradication was so complete that the term *saloon* would not be used again in

Atlanta until the 1960s, and then only sparingly. Even historical photographs of Atlanta saloons are rare. Today, the term is still not popular, unless a western theme is involved.

The old drinking establishment was replaced by a mail-order business that was neither personal nor quintessentially Atlanta. The Atlanta Brewing and Ice Company closed its tied houses and removed the fixtures for resale. The company amended its charter to reflect the new name Atlanta Ice and Bottling Company and sought to produce "soft" drinks with less than 0.5 percent alcohol. These actions were taken to avoid a lawsuit from the Georgia Anti-Saloon League, which might pursue the sales of near beer and violations of the anti-advertising elements of the new law. As the new name indicated, Atlanta Ice and Bottling used existing refrigeration equipment to produce ice for sale. Many breweries in Georgia became bottling interests. The Acme Brewing Company in Macon became a meatpacking plant; others simply shut their doors.

The anti-advertising law forbade advertisements for alcohol in any media format within Georgia. Dry lawmakers had learned their lessons from earlier loopholes. The law was written with exhaustive detail on every form of alcohol including its manufacture, storage, transportation and consumption. The only legal consumption was of minor personal purchases via delivery from unadvertised sources (but no Sunday deliveries). Beer ads in Atlanta newspapers prior to the May 1 advertisement ban urged Atlantans to cut out the listings, complete with ordering instructions and price lists, and keep them handy for "a long Georgia Sahara." Other ads promised lager beer at less than one cent per glass delivered via parcel post. An ad in late April 1916 read, "BEER MADE AT HOME: New Discovery Benefits Thousands."

As soon as the May 1 deadline passed, the ASL and prohibition-minded Georgia politicians redirected their attention to private stocks of alcohol. On May 2, Georgia Superior Court judge Ben Hill declared to the *Atlanta Constitution* that anyone with a private supply of alcohol beyond the limit was an anarchist. In the same newspaper, an agent of the ASL was quoted as saying, "Our agents have information and evidence that a number of prominent citizens have stored large quantities of whisky and beer in the basements of their homes." The ASL was on the prowl and wanted to bring down private liquor. Homeowners argued that any alcohol purchased prior to May 1 was not subject to the current law and therefore legal. The ASL didn't care for public opinion.

On May 8, 1916, the Atlanta police raided a house on Barwell Road and found a system of tunnels, a "drinking" well packed full of distilled spirits and an attic stuffed with over two hundred bottles of beer. The renter of the property, Jim Daulis, had been arrested several nights before at his Greek restaurant at 199 Peters Street, where 200 pints and 281 half pints of distilled spirits and a barrel of beer were seized. In total, Mr. Daulis was arrested four times at different locations in the same month.

Atlanta police would make many similar raids. Despite police efforts to control blind tigers and illegal selling of alcohol, the problems persisted from 1916 through 1918. In a December 1918 *Atlanta Constitution* article titled "Pistol Toting and Whisky Drinking," the newspaper states there had been 482 cases for prohibition violations from January to November 1918. There were 315 such cases in the same period in 1917. As to drunkenness, 2,077 cases were reported for the first eleven months of 1918; 2,028 cases were noted for that period in 1917.

NATIONAL PROHIBITION

By this time, the United States' entering the growing conflict in Europe seemed likely, and the federal government began looking for ways to fund the war effort. From 1868 to 1913, 90 percent of all government revenue came from the sale of beer, wine, liquor and tobacco. The Anti-Saloon League saw an opportunity to weaken the economic argument for keeping the alcohol industry alive. The federal income tax, if reinstated, would provide revenue from an alternate source. The ASL threw the full weight of its support and lobbying efforts behind the passing of an income tax mandate. In 1913, the Sixteenth Amendment went into effect (and has been the topic of ongoing political debate since). This proved both the lobbying power of the ASL and its strength to influence Constitutional law.

With the country entering World War I in 1917, anything with a German connection was perceived as unpatriotic. Sauerkraut became "Liberty Cabbage," and all things German, including lager beer, became taboo to many people. The ASL used the fact that most brewers in America were of German ancestry to stereotype them as anti-American. State after state went dry. New national laws reduced the breweries to shadows of their former selves. National Prohibition—once viewed as a pipe dream by not only the American people but also many in the temperance movement—was a foregone conclusion. Drys had played their cards well and brought the existing American political system, via the ASL, in line with their agenda.

On the eve of national Prohibition, a person dressed as Uncle Sam rode a symbolic water wagon around Atlanta letting everyone know of the big celebration planned that night. A biplane flew low over the city and dropped flyers announcing the festivities, courtesy of the Georgia ASL. As night descended, state temperance leaders gathered in a congratulatory meeting and then joined a parade headed for downtown. At midnight in the center of Atlanta's Five Points, a large effigy of John Barleycorn (the personification of alcohol) was burned in mock sacrifice. Contraband booze was poured onto the burning figure as the crowd cheered. Enacted by the Eighteenth Amendment of the Constitution of the United States, the National Prohibition Act (Volstead Act) went into effect on January 17, 1920. The United States was declared legally dry.

It didn't take long for the American public to call foul. Within a month, a national proposal for an amendment to allow for "light wines and beer" was proposed, though it was ultimately shot down. In response, America became a nation of scofflaws. The term *scofflaw* arose from a 1924 *Boston Herald* competition to describe a person who defied the law by drinking illegally.

Apparently, alcohol was more a social issue than the perceived moral issue. Drys watched in horror as their dream became a national nightmare. The Volstead Act called for joint law enforcement at federal, state, county and municipality levels. Even with all these enforcers involved, illegal alcohol use persisted. Prohibition at a national level did not solve what the temperance movement reproached as the nation's obsession with alcohol any more than local option and state prohibition did. The majority of American voters, it seems, was "moist" (moderate imbibers of their choice of ethyl alcohol).

By 1933, the United States had endured enough of this issue and its inherent problems. The Twenty-first Amendment of the Constitution was ratified, abolishing the Eighteenth Amendment and the entire concept of Prohibition and temperance. The State of Georgia, however, was not ready to separate from its temperance roots and has never officially ratified the Twenty-first Amendment. Shortly after the adoption of national repeal, Atlanta—using local option—authorized the sale of 3.2 percent ABV beers. Forty-nine establishments were licensed the first day, and beer sold so fast bartenders did not have time to chill it.

In 1935, Georgia attempted to repeal its state prohibition but failed due to conservative votes. Spirituous liquors were still forbidden in Georgia, but amendments were passed making wine and beer legal throughout the state, with beer set to an upper limit of 6 percent ABV.

Later that year (1935), Governor Eugene Talmadge signed legislation providing for the licensing and taxing of beer and other malt beverages. The State Revenue Commission, which administered the law, provided stamps to retailers to document that the state tax had been paid on each container. In 1938, the Georgia General Assembly passed the local option on liquor, and most urban counties became fully wet. The new state law required customers to sign their names when making alcohol purchases. Many people chose to sign as the governor or the revenue commissioner.

Atlanta's road through temperance and prohibition was long and winding. The journey started with a brief dry period from 1886 to 1888, followed by varying state prohibition laws from 1908, then national Prohibition (1917–1933) and finally state prohibition ending in 1938. The end result is a total of thirty-two years of prohibition, thirty of which were consecutive. Luckily, the Atlantic Company would form from the remnants of the former Atlanta Ice and Bottling Company to carry on the city's early brewing traditions after state and federal prohibitions ended.

LIFE AFTER PROHIBITION

There was no clear winner in the American Prohibition battle. The separation of church and state became dangerously thin during this period. The battle led to the birth of pressure politics and demonstrated the dangers of defining moral and social statutes for a diverse population by a select few. Legal alcohol consumption resumed, but America's perceptions of alcohol were changed forever. The temperance concept that any and all alcohol and its use are inherently evil still exists in segments of American culture.

Some of the prohibition laws that were enacted as far back as the 1880s are still active today as blue laws. Only recently has Georgia seen the change in the legal ABV limit of beer, allowing craft and other specialty beers to be sold. The sale of alcohol on Sunday—finally allowed in many cities and counties across Georgia by a vote in 2011—is still a contentious issue in the Atlanta metropolitan area, 126 years after it first appeared on city ordinances.

Local option was originally a means of using the power of rural voters to legally mandate prohibition over wet cities. Local option later became the reason for the wide variety of dry, wet and moist counties in Georgia.

In an attempt to control monopolies (via tied houses, distillery- or brewery-owned or operated saloons), the federal government created a three-tiered system of alcohol regulation and passed most of the control to the individual states. The system was also intended to maintain government regulation and taxation of alcoholic beverages. Each state has the power to determine how it wishes to organize the three-tier system. The three levels in the hierarchy are producers (brewers, distillers and winemakers), wholesale distributors and retailers (bars, package stores, grocery stores and so forth). Unfortunately, the attempt to limit the producers' putting pressure on retailers did not work exactly as planned. (A person could write a book on this subject. Macrobreweries continue to achieve the widest distribution and occupy the vast majority of taps.)

The current exception to the three-tier rule is the brewpub, which eliminates the distribution tier for beer both brewed and consumed on premises. As of early 2013, proposed Georgia House Bill 314 and Senate Bill 174 would allow Georgians to buy limited quantities of beer directly from the brewery or brewpub as a take-home item. It remains to be seen whether these laws exceed the tolerance of Georgia's lawmakers.

Georgia is painstakingly moving toward a more modern alcohol-regulation system that balances between the social aspects of the adult-beverage industry and government control. With the softening of some of the regulations within the three-tier system, Atlanta now boasts modern brewpubs, beer tastings at events and growler shops that fill containers for patrons to take home. The changing climate is again supporting small, local breweries so common in earlier America.

CHAPTER FOUR

FULL OF GOOD CHEER

The Atlantic Brewery Story

Don't tell me about the beer of the "good old days," I made it and I drank it.
—*Alois J. Reis*
Head brew master of the Atlantic Brewery
Atlanta Constitution, *1943*

As we mentioned previously, Dionis Fechter came with his brother Egidius to Atlanta in the early 1850s from the southwestern German state of Baden-Wurttemberg. The brothers brought with them the technology and German tradition of brewing lager beer. Egidius opened Atlanta's first brewery in Howell Park, and by 1858, his brother Dionis was in the brewing business with O. Kontz at the City Brewery on the Western and Atlantic Railroad track near the intersection of Marietta and Pine Streets.

Unfortunately for Kontz and Fechter, the Civil War began a few years after they opened their brewing operation. In 1864, the city of Atlanta underwent the "Sherman Renovation Plan" (that is, the city was burned by the U.S. Army), which put most city operations on a bit of a hiatus. Although the City Brewery survived the burning and resumed operations in 1865, Kontz was bought out that same year by new investors Michael Kries and F.A. DeGeorgis. Porter-style ale was added to lager as the brewery's two beer offerings, and the name briefly changed to the "Old City Brewery," which produced roughly one thousand barrels of beer per year (an estimated thirty-one thousand U.S. gallons based on the contemporary thirty-one-gallon barrel).

Estimated location of the original City Brewery, modern-day Northwest Marietta Street. *Ron Smith.*

In a bewildering change of owners, investors and structures that confused consumers and even the local press, Michael Kries would leave to open the Fulton Brewery in 1866. Edward Mercer would emerge to partner with Dionis Fechter, unveiling a newly combined company interest by 1871: Fechter and Mercer's Atlanta City Brewing Company. A new and larger brewery was built, complete with lagering cellars, at the corner of Harris and Collins Streets (modern-day Courtland Street and John Portman Boulevard). Its ten employees tirelessly produced lager and ale year-round and sold its spent grain to local dairy farms. The brewery's new beer lineup included four beers: "Genuine Lager Beer, The Celebrated Southern, XXX Ale and the Premium Beer." The July 27, 1873 *Macon Telegraph and Messenger* mentions that one of the Atlanta City Brewery's beers had an alcohol by volume of 6.6025 percent and a specific gravity of 1.0158—quite a detailed analysis for 1871.

The years between 1874 and 1876 witnessed major changes for the brewery. Dionis Fechter passed away on February 3, 1874; his brother Egidius became the administrator of his estate and his holdings in the brewery. By 1875, an ice machine built by the Columbus Iron Works had been added to the brewery, introducing a modern marvel that produced ice from chilled steel plates cooled by gaseous ammonia. The brewery had three horses that pulled flatbed wagons to deliver one hundred kegs of beer daily to the thirsty citizens of Atlanta. By this time, the Atlanta City Brewery was one of the largest brewing interests in the Southeast, but competition from the distant giants in Milwaukee and Cincinnati was always present.

Incorporation and Expansion

In February 1876, Egidius Fechter, W.H. Tuller, A.J. Kuhn, H. Werner and Joseph Fleishel incorporated the Atlanta City Brewery with a capital stock value of $50,000. At the time, the brewery covered nearly four acres of land in Atlanta. Like all larger breweries of the time, the Atlanta brewery was powered by coal. Large coal-fired steam engines drove the machinery within the brewery. Large smokestacks streamed coal smoke into the air as the engines drove gears, pulleys and belts to run pumps for the refrigeration machinery and to move the developing beer through its stages. Forty tons of coal mined from Coal Creek, Tennessee, were needed to run the brewery each month. To the workers, there was an ever-present danger of fire and the potential of an exploding steam boiler. The Atlanta City Brewery caught fire twice. In 1874, a fire partially burned the brewery, and in 1880, the brewery completely burned down, leaving only the cellars. The new 1880 brewery was built completely of brick and stone to reduce fire risk.

Theodore Fechter, a member of the originating Fechter family, was the brewer for the Atlanta City Brewery in the late 1880s. Hops for his brewing were shipped from Bohemia (Czech Republic area), Bavaria (southern Germany) and California, while the malted grain originated from the Kentucky Malting Company in Louisville. A portion of the beer was placed in barrels of various sizes, which had already been washed and steam cleaned for reuse. The remaining beer was bottled at the rate of 2,500 bottles per day. Until January 1889, the bottling was managed by the Southern Bottling Company but was then consolidated into the brewery's internal operations. The brewery had sixty-five retailers in the city (often tied houses) and regionally distributed beer in all directions, covering large areas of Georgia, Alabama, North Carolina, South Carolina and Florida.

The Atlanta City Brewery produced varying types of lager beer between 1875 and 1890. The brewery's beers were detailed by the *Columbus Daily Enquirer* in 1877 and the *Macon Telegraph and Messenger* in 1875. Brews

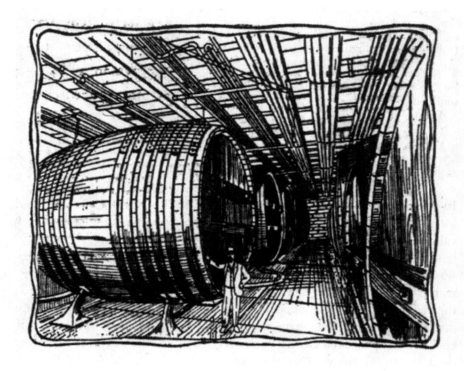

The massive wooden vessels in the Atlanta Brewing and Ice Company cellars. Astoria beer ad in the *Atlanta Constitution*, July 11, 1903. *Image courtesy of www.Fold3.com.*

included a "Champagne Lager Beer" and a seasonal maibock (May bock, based on the time of year it was typically brewed). A Macon, Georgia journalist—uncertain how to spell or pronounce the beer—called the maibock a "Buck Beer."

Immense lagering cellars were a prominent feature of the Atlanta City Brewery. The cellars are described as having arched, cathedral-like ceilings made of solid masonry. The brewery's cellars also contained a well. The water from this well was pumped day and night from the underground spring at the reported rate of 1,000 gallons per hour for use in brewing, cleaning, steam production and making ice. There were nine beer vaults, which extended thirty-five to forty-five feet below street level. The brewery's cellars held huge wooden vessels with a capacity of 1,000 to 2,200 gallons each for fermenting, storing and clarifying. These large wooden vats would play an important part in the beer throughout the brewery's existence. Atlanta City Brewery aged all of its beer in wooden containers. The lagers were aged an exceptionally long time—sometimes through an entire season. In 1878, the beer stock on hand was typically four thousand barrels and kept growing year by year to hundreds of thousands of gallons.

To keep the beer cold, on site or in transport, the brewery's ice production was impressive. At first, massive blocks of ice were continuously placed around the lagering barrels to keep the beer chilled. The runoff from the slowly melting ice would have made the cellar damp and difficult for maintaining a consistent temperature—a less-than-ideal storage facility. Later, pipes from an ammonia refrigerating system were used to keep the cellars cold and did so without the excess moisture that caused the wood of the barrels to rot prematurely. By 1904, the brewery could maintain a temperature variance of one degree Fahrenheit by using the chilled pipes. The presence of the cellars by 1871 and descriptions detailing the handling of the beer during production show that the beer was "fully lagered" as opposed to the quickly produced and consumed schenk beer that was common in earlier practices.

Atlanta Brewing and Ice Company

By the time of Edward Mercer's death in 1885, both Mercer and Egidius Fechter had sold their interests in the brewery. This financial move might have been prompted by the rising temperance sentiment in the Atlanta area. The change in the company name in 1891 to the Atlanta Brewing and Ice Company might also have

Albert Steiner, president of the Atlanta Brewing and Ice Company for more than nineteen years. Upon his death in 1919, he donated millions of dollars to build the Steiner Cancer Center (Grady Health System). *Photo courtesy of the Cuba Archives of the Breman Museum.*

been a response to the times. The brewery was producing a large amount of ice that could be sold independently of brewery operation and sales if the need suddenly arose (as it might under alcohol prohibition). Ads in the *Atlanta Constitution* in 1896 show the brewery as producing the cheapest ice sold in Atlanta. In 1899, the Atlanta Brewing and Ice Company warned the *American Brewers' Review,* "Gentlemen…we think it is a good idea to abandon the publication of the output of breweries in your annual breweries guide—Very respectfully." This caution is clearly in response to the temperance movement's tactic of using brewery statistics to promote its cause.

H.G. Kuhrt was president of the brewery until the late 1880s, handing the reins to Mr. C. Beerman (a perfect name for the brewing business), who served in the role from 1890 to 1897. Files held by the Atlanta History Center show that Beerman and Kuhrt also owned and operated a thriving "segar store" in Atlanta. Albert Steiner, a Central European Jewish immigrant, became president of the Atlanta Brewing and Ice Company in 1897, a position he held until the mid- to late 1910s. (One of the brewery's longest-produced beers, Steinerbräu/Steinerbru, was named after him.) Steiner brought a

Detail of an Atlanta Brewing and Ice Company Old Cabinet export lager stoneware bottle. Note the company emblem, barley and cherub spilling a basket of hops. *Bottle available for photographing courtesy of Lee Connell and Tony Carr Jr. Ron Smith.*

The Atlanta Brewing and Ice Company, circa 1900. *Photo courtesy of the Kenan Research Center at the Atlanta History Center.*

new assistant brew master over from Bohemia in 1901. This brewer, John Bips, stayed with the brewery for over forty years. Head brew master Alois J. Reis, already with Atlanta Brewing and Ice Company, also stayed with the company for decades.

The brewery had expanded its distributing business during the restrictive Fulton County prohibition (1886–1888), and by the late 1800s, a large portion of the beer was sold to the brewery's expanded network. The export beer was either "bottled and boxed" or barreled under two names—Old Cabinet Lager or Southern Export Select Lager Beer—and distributed to the southeastern United States. In some cases, the beer was shipped out of the country.

By 1903, the output of the brewery had grown to over seventy-five thousand barrels per year distributed across twelve states. The brewing facility had expanded to eighty thousand square feet and covered another acre of property (totaling five acres). The brewery had increased malt and hops storage, a refrigerating plant producing 250 tons of ice per day, a cooperage (barrel-making station), a blacksmith shop, 125 mules and horses and a fleet of wagons. The large storage and fermenting vat capacities ranged from 1,500 to 5,000 gallons each. The cellars were enlarged to hold a total of 125 of the giant casks. Many miles of piping existed within the brewery to convey the beer from boilers to vats, to cooling casks, to filters and, ultimately, to the bottling house.

The brewery's bottling department filled more than twenty thousand bottles per fifteen-hour workday. The bottles were corked and caged, spaced

Drawing of an early 1900s Astoria beer bottle, shown with glass and Atlanta Ice and Bottling Company corkscrew. Drawing by Rose Dunning, 2013. *Courtesy of Rose Dunning*

and separated by corrugated cones and crated for shipment. Between two and three hundred men were employed at the brewery, producing Old Cabinet Lager, Royal Pale and the brewery's new product—Astoria beer. The new brand was grandly advertised as being made with fresh Bohemian hops brought by ships in the Hamburg-American line, first to the port of New Orleans and then by train to Atlanta.

PROHIBITION AND THE ATLANTA ICE AND BOTTLING COMPANY

With Georgia becoming a dry state in 1908, only nonintoxicating beverages could be legally sold. The Atlanta Brewing and Ice Company would file a statement, as required by law, that it would only "manufacture, deal in and sell any and all kinds of non-intoxicating drinks." The classification nonintoxicating would include near beer of no more than 4 percent ABV. The brewery would continue to produce near beer, keeping a low profile to avoid providing any information to the local Anti-Saloon League and Woman's Christian Temperance Union. The only advertisement and labels that we could locate for the time period were for Royal Pale Beer, Bavarian German Style Beer and Pilsener German Style Beer. The Royal Pale Beer label notes, "Contains About 4% Alcohol."

The 1916 amendments to the Georgia prohibition law had adverse effects on the Atlanta Brewing and Ice Company. The anti-advertising portion of the law forbade advertisements for alcohol in any media format within Georgia, including business names containing the words brewery or brewing. To avoid a lawsuit by the Georgia Anti-Saloon League, the Atlanta Brewing and Ice Company changed its name to the Atlanta Ice and Bottling Company and amended its charter to reflect the new name. The evolving company stayed in business by producing "soft" drinks with less than 0.5 percent alcohol to avoid legal action. Although this change was extreme, it staged the company to be in compliance when the National Prohibition Act (Volstead Act) went into effect in January 1920. Evidence also exists that the Atlanta Ice and Bottling Company bottled soda-based drinks for a large local cola company.

The Atlanta Ice and Bottling Company produced at least four nonalcoholic cereal beverage products during Georgia prohibition and into national Prohibition: Royal, possibly a prohibition version of its earlier Royal Pale

Atlanta Ice and Bottling Company Prohibition-era Veribest cereal beverage label, circa 1920s. *Image courtesy of Ken Jones.*

Beer, with an ABV of 0.47 percent; Veribest, with a similarly low ABV; and two beverages—Würzburger Style and Original Swiss Brew—for the local distributors of Marshall & Reynolds.

RISE OF THE ATLANTIC BREWERY

The Atlantic Ice and Coal Company, a competitor of the Atlanta Ice and Bottling Company since 1910, built a brewery in Chattanooga, Tennessee, under the name Southeastern Brewing Company in August 1933 (shortly before the repeal of national Prohibition). It began brewing in 1934 under the product name Old South, distributing this beer in Tennessee, Georgia, North Carolina, South Carolina, Virginia, Florida and Alabama. In 1935, due to copyright entanglements, the product name was phased out and the Atlantic beer line was born.

In the same year, the Atlantic Ice and Coal Company purchased the Atlanta Ice and Bottling Company and the rights to its former beer lines. The Atlanta Ice and Bottling Company had just begun to brew beer after the repeal of Georgia prohibition. Under the Atlanta Ice and Bottling Company the Steinerbräu line had been resurrected as Steinerbru. Once acquired by the Atlanta Ice and Coal Company, the brewery continued producing this brand in both "beer" (a pilsner), ale and its seasonal bock. These three styles were also available in its new Atlantic line, with the labels announcing Atlantic Sparkling Ale and Good Old Atlantic Beer and sporting a goat's head on all its bock beers. The standard four-beer lineup was described in a 1943 *Atlanta Constitution* article, with Atlantic Beer as a "yellow gold Pilsener," Atlantic Ale as having a "stouter touch" of the same flavor found in Atlantic Beer, Steinerbru Beer as a "darker beer with a 'winey tang'" and Steinerbru Ale as being "rich bodied and commanding."

The Atlantic Ice and Coal Company breweries adopted the name of the beer lines, becoming the Atlantic Company. Each local brewery was called "Atlantic Brewery." During the first decade after Prohibition, the Atlanta location produced Belle of Georgia for Beverage Distributors, Inc. in Augusta. This beer was a pre-Prohibition favorite among the citizens of Augusta and the surrounding area but had become unavailable due to the closure of the Augusta Brewing Company.

The Atlantic Company would ultimately buy or build five breweries: Chattanooga, Tennessee (1933); Atlanta, Georgia (1935); Norfolk, Virginia (1936); Charlotte, North Carolina (1936); and Orlando, Florida (1937). Beer was produced and distributed in kegs, bottles and cans. Flat-top beer cans were on the market by 1935. However, many breweries would continue to use cone-top cans for many years because this type of can could run through existing bottling machinery.

By 1939, the popular slogan "Full of Good Cheer" was used in association with Atlantic Beer. In 1940, another brand, Signal Pale, was introduced to the market. This beer was likely named in the company's Chattanooga market for the railroad industry's use of signal lanterns to communicate among train workers. A second hypothesis relates to the signals passed from Signal Mountain to troops during the Civil War.

In the late 1940s, the Atlantic Brewery started using a water treatment process it referred to as Schweitzer. The labels of most of the beers in its product line boasted, "Now Schweitzer-ized Secret Swiss Process." Apparently, this process was similar to what is presently called Burtonisation (matching the brewing water to that of the River Trent in England, a preferred water source). This process is used when a brewer wishes to accent the hops in a light or pale beer. The trademarked process name soon began appearing not only on Atlantic Brewery beers but also on the delivery trucks that had replaced the horse-drawn drays used at the turn of the century. The

The Atlantic Brewery in 1950. Lane Brothers Collection. *Courtesy of the Special Collections and Archives, Georgia State University Library.*

fleet of trucks included a variety of vehicles, typically white, emblazoned with Atlantic Beer, Steinerbru Beer or Ale and the now familiar Schweitzer-ized logo.

An amazing 120 million bottles of beer were sold by the Atlantic Brewery in 1942. This statistic no doubt includes substantial sales of Atlantic Beer to the U.S. Army and Navy during the early part of World War II. This year was most likely the pinnacle of the Atlanta facility's production. Ironically, World War II was also very hard on the Atlantic Company with the rationing of tires, gasoline, available trucks and containers. An article in the January 8, 1944 *Billboard* indicates that for a short period, the company might have stopped brewing altogether. In the same year, a bizarre fire in the Atlanta brewery's hops storage area destroyed $100,000 worth of hops, setting the brewery back even further.

THE FINAL YEARS

Records from 1944 to 1955 are scarce for the Atlanta location of the Atlantic Company. Business as usual seems to occur, but the U.S. media was focused on the end of World War II and the Korean Conflict. What is known is that

The Downtown Atlanta Hilton Hotel at the former site of the Atlantic Brewery. Many travelers park in the underground garage not knowing the garage replaced the cellars where the former Atlantic Brewery lagered (stored) its beer. *Ron Smith.*

The Heart of Atlanta Hotel did not survive long. The hotel was torn down by the same company that leveled the Atlantic Brewery. This time the demolition was to build the Hilton Hotel in 1973. Surprisingly, the massive cellars of the former brewery still existed under the Heart of Atlanta Hotel. A reporter walked the cellars along with the president of the demolition company. The reporter commented on the stone cellars and the well, which contained crystal-clear water. The stone-lined cellars were removed to make space for the parking garage of the new downtown Hilton Hotel.

the Steinerbru Room, a tasting and entertaining room at the Atlantic Brewery, was maintained as late as 1954. This was Atlanta's first documented brewery sampling room. The room was used to promote products to local and traveling VIPs at a time when a few breweries were becoming the predominant producers of American beer.

The Atlantic Brewery closed its doors in 1955. The brewery's attention to detail and flavor might have hastened its demise. Its long aging process and more flavorful beers placed a higher burden on the brewery than did other breweries' technologies for mass production. The American palate had changed through Prohibition; the common beer was now a rice and corn adjunct pilsner that was a shadow of its Bohemian forefather.

The brewery was sold for $250,000, and the buildings were promptly leveled to build the Heart of Atlanta Hotel. The loss of the Atlantic Brewery tolled the death of a brewing business that spanned nearly one hundred years in Atlanta. It also began a four-decade period during which no locally owned and operated brewery existed in the Atlanta area. A national trend toward large-scale (macro) breweries had begun, and for many years, a handful of breweries would produce the majority of domestic beer. In 1955, there were roughly 250 breweries in America, and this number was rapidly declining.

CHAPTER FIVE

CARLING, COLD CUTS AND PSEUDO-SALOONS

This ain't no beer joint. It's a community center.
—Manuel Maloof
Owner of Manuel's Tavern, quoted in the Day, *1986*

ATLANTA'S CARLING BREWERY

In 1958, the Canadian-based Carling Brewing Company opened a brewery in Atlanta at a cost of $15 million. This brewery was opened on the heels of an aggressive marketing campaign led by the company in the southern states. Plans for the Atlanta brewery had been laid ten years earlier, when the company expanded its market to the area. Five hundred local political leaders and businessmen, including the mayor of Atlanta, were dined and led on a tour of the new brewing facility. The president of Carling at the time, Ian R. Dowie, was on hand for the opening ceremonies. It was touted as being the "first brewery to be built in the South in more than 25 years" and "the world's most modern brewery."

The new brewery had a capacity of 350,000 barrels a year, and the canning line filled six hundred cans a minute. The signature Carling's Black Label was produced in Atlanta, along with Red Cap Ale, Heidelberg Light Pilsener and Calgary Malt Liquor. The plant employed around one hundred employees. Carling distributed beer from the operation to nine southern states through a network of seventy-two distributors. From the 1950s to the

The former Atlanta Carling Brewery building. Located on I-75, south of Downtown Atlanta. *Ron Smith.*

'60s, the brewery had a large parking lot, and the Atlanta Kennel Club held its dog shows in the lot twice a year.

Carling Brewing Company saw sales peak in the 1960s, but in the next decade, sales began slipping. Better refrigeration technology and the Eisenhower Interstate System of the 1950s had helped the mega-breweries move their products faster across America. This allowed the big producers to compete directly with every local and regional brewery. The large breweries could also afford to advertise in a new medium—television—using catchy jingles and slogans to bring their product names to every corner of the country.

The company was not only out-advertised by the larger breweries but also could not keep manufacturing costs low enough to price its beer competitively. In 1972, Carling sold the Atlanta brewery facilities to Coca-Cola Enterprises for use in soft drink production. The old Carling Brewery building still stands and can be seen from I-75. With the closure of the Carling Brewery, no locally produced beer was available in Atlanta once again. Large national brands, like Budweiser, Pabst, Blatz, Miller, Schlitz, Cooks and Schaefer, streamed into the Southeast, filling the gap.

DELICATESSEN TO TAVERN:
DO YOU WANT A SANDWICH WITH THAT BEER?

Prior to Prohibition, beer was mainly consumed from draft in a tavern or saloon. After Prohibition, improved bottling and canning technology allowed beer to be more portable. Many taverns, restaurants and delicatessens began using canned and bottled beer, which could be stored in their existing refrigerated systems without their having to buy expensive draft lines and add keg storage.

Residents of Atlanta did not stop drinking beer when it was no longer locally produced. Places like Atkins Park Tavern, Manuel's Tavern, Northside Tavern, the Royal Peacock, Underground Atlanta and Red Dog Saloon kept beer flowing to thirsty Atlantans. Some were established entertainment venues offering beer or a full bar to liven the night's social interaction. Many began as delicatessens that served beer in order to offer a complete meal. They were also the only places in town that had a bar-like counter for seating.

Manuel's Tavern

Harry's Delicatessen, a sandwich shop and beer joint on the corner of North Avenue and North Highland Avenue, was purchased in 1956 by Manuel Maloof. A World War II veteran, Maloof wanted to create a local tavern that resembled the pubs he saw in England during the war. The wood bar in Manuel's Tavern originated from his father's establishment, the Tip Top Billiard Parlor on Pryor Street, which opened in 1907. This one-hundred-year-old bar is one of the few remaining artifacts from Atlanta's saloon era.

For many years, Manuel's followed Harry's model and operated solely as a sandwich shop and beer bar. The bestselling beer brands were Andeker, an all-malt Munich-style lager by Pabst Brewing Company, and Budweiser. Because of the tavern's proximity to then-dry DeKalb County, students and faculty from Agnes Scott College and Emory University would come across the county border to meet their friends for lunch, dinner and drinks. The bar became known as the intellectual and freethinking hot spot of the area. Later, this freethinking and drinking would morph into a political climate that paralleled the political career of its owner, a liberal Democrat.

What ultimately evolved was an Atlanta icon. Manuel's Tavern is a colorful place in every sense of the word. Its décor is assorted, and its patrons are from every walk of Atlanta life. Loyal customers over the years have included day workers, contractors, writers, sports figures, philosophers, Atlanta mayors and U.S. presidents—collectively draining a weekly average of seventy half-barrels of beer.

The walls speak to tell the story of the surrounding community. The tavern displays nearly sixty years of mementoes from the family and patrons of the establishment. The mix includes beer cans, photos of prominent political figures, a painting (female nude) given as payment for a bar tab and the ashes of Manuel Maloof and another beloved patron. No other establishment in Atlanta encapsulates the city quite like historic Manuel's Tavern.

Moe's and Joe's

Moe's and Joe's has served the Virginia-Highland community of Atlanta since 1947. *Ron Smith.*

In 1947, brothers Moe and Joe Krinsky opened a tavern in the Virginia-Highland district of Atlanta, serving cold beer, hot dogs and hamburgers. Like fellow tavern owner Manuel Maloof, the brothers benefited from having their establishment close to the border of then-dry DeKalb County. Ambiance and long-term employees giving great service helped immensely.

In the mid-1950s the tavern became a voluntary "tied house" for Pabst Blue Ribbon (PBR) and became famous for its support of the "Finest Beer Served." Although Moe, Joe and their family are no longer owners of the business, little has changed over the years (although the beer list goes beyond PBR). Moe's and Joe's remains one of Atlanta's legendary spots.

Atkins Park Restaurant and Tavern

The historic Atkins Park community started out as a streetcar suburb of Atlanta. Within this area, now a part of Virginia-Highland, is Atkins Park Restaurant and Tavern. The establishment opened circa 1921 as a delicatessen run by cousins Morris Franco and Reuben Piha. Atkins Park Tavern claims to be the oldest licensed alcohol-serving establishment in post-Prohibition Atlanta, applying for its license at the repeal of Georgia prohibition.

Operating as a deli until the late 1970s or early 1980s, Atkins Park Tavern served a large variety of cold sandwiches, boiled and pickled eggs and dill pickles. To wash the sandwiches down, customers could choose from two beer taps and a variety of beer in cans and bottles. Local beers in the 1960s were limited to Carling Black Label and Red Cap Ale from the nearby Atlanta Carling Brewery. National brands like Blatz, Budweiser, Bush, Falstaff, Miller, Pabst and Schlitz were more common. Imports were limited to Heineken and Löwenbräu. Draft beers were twenty-five cents for ten ounces in the late 1960s and were poured into pilsner-style glasses or small mugs.

Opening in the 1920s, Atkins Park Tavern in Virginia-Highland is one of the oldest bars in Atlanta. *Ron Smith.*

From the 1950s to the 1970s, Atkins Park had a large population of blue-collar workers who stayed in boardinghouses during the week and returned to their distant homes (mainly in North Georgia) on the weekends. Many of these workers ate and drank at the tavern. Georgia Tech students also frequented the deli. Residents from the nearby dry county of DeKalb would cross the border to frequent Atkins Park, along with Moe's and Joe's and Manuel's Tavern.

According to previous owners, Atkins Park was unique among deli-taverns in that it offered beer and wine to go for many of its operational years. Wine, however, could not be imbibed on site, since the tavern did not have an on-premise wine license.

The neighboring space (a dry cleaner) was annexed in the late 1970s, becoming a pool hall attached to the original barroom section. This addition would ultimately become a restaurant as Atkins Park Tavern became a combination tavern and restaurant. In 1983, Warren Bruno (also of the former Phoenix Brewpub and Ormsby's) purchased the location. In its lifespan of over ninety years of operation, Atkins Park Tavern has had only five owner/operators as of this writing. The Atkins Park Tavern Restaurant Group, conceived by Bruno, currently operates the original Atkins Park Tavern location.

THE RETURN OF THE SALOON! WELL, SORTA

Red Dog Saloon

Whether by knowledge or accident, the saloon did somewhat return to Atlanta. One of the first contemporary saloons, in name and true 1800s style, was the Red Dog Saloon on Roswell Road in Buckhead. The interior was reminiscent of an earlier saloon, with wooden bar and accents, brass rail, no bar stools and female nude décor. The saloon even featured doors on two intersecting streets (Prohibition style). Several beer taps supplied the house draft but were stocked with beer produced out of town.

The Red Dog Saloon in 1961, Buckhead area of Atlanta. *Photo by Tracy O'Neal. Courtesy of the Special Collections and Archives, Georgia State University Library.*

Underground Atlanta

In the 1970s, the Underground was a national hot spot with world-famous stars and rock bands making appearances in the venue. In the center of this historic hubbub was Muhlenbrink's Saloon, named in honor of Hans Muhlenbrink, noted Atlanta saloon owner and liquor wholesaler from the late nineteenth and early twentieth centuries (see chapter two).

The legendary William Lee "Piano Red" Perryman played piano for a decade at Muhlenbrink's Saloon. He played in a manner that propelled him through time and musical style, including ragtime, jazz, blues and rock-n-roll. His timelessness captivated all audiences, including the Rolling Stones. His song "Underground Atlanta" hails the virtues of the time in the city. However, Piano Red wasn't the first noted musician to sing about the place.

In the 1920s, the City of Atlanta built a series of viaducts to move street traffic above the railroad tracks located downtown. Pedestrian, horse-drawn and

Peachtree Street Viaduct, Raphael Tuck & Sons postcard series number 2449, circa early 1900s. *From the personal collection of Ron Smith.*

motor vehicle traffic interactions with trains posed ongoing risks. In response to the elevated street level, businesses moved their storefronts up one floor. This created a hidden world where the majority of the public rarely ventured. This underground area became a prime spot for nightclubs that functioned as blind tigers during state and national prohibitions. The song "Preachin' the Blues" by Bessie Smith—a blues songwriter, singer and all-around legend—proclaims, "Down in Atlanta G-A under the viaduct every day drinkin' corn and hollerin' hooray, pianos playin' 'til the break of day."

Muhlenbrink's Saloon was not the only Underground Atlanta nightspot to sling beers and tunes. The Blarney Stone billed itself as the "world's largest and Atlanta's only authentic Irish Pub." Ruby Red's Warehouse, a club based on the Roaring Twenties, promised banjos plus booze plus beer plus fun where "waiters in fancy vests and sleeve garters serve beer and beverages."

Another Underground Atlanta venue named after an 1800s Atlanta icon was P.J. Kenny's Saloon. Although not as famous as Muhlenbrink's, the saloon strove to resurrect an earlier period in Atlanta history that was filled with wild times, free-flowing beer and spirituous liquors.

The word saloon was nearly outlawed in America, thanks to the Anti-Saloon League and the combined forces driving the temperance movement. Even within the repeal process of Prohibition, President Franklin D. Roosevelt stated, "I ask especially that no State shall by law or otherwise authorize the return of the saloon either in its old form or in some modern guise." He need not have worried; the majority of the American public ultimately forgot the saloon outside the Old West legends.

FROM HOMEBREWING TO CRAFT BREWING

By the 1970s, light lager brewed with added rice or corn remained the only viable beer style produced in the United States. Low-calorie versions of this style also began to grow in popularity as the country became more health conscious. What little import beer sales did occur originated primarily from Canada, Mexico, Australia, Germany and the Netherlands. Topping the import sales was Heineken. During the 1970s, Carling produced the Danish Tuborg and Miller brewed a modified German Löwenbrau, with both domestically produced beers marketed as premium imports.

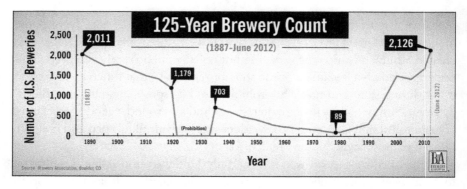

The 125-Year Brewery Count. By the late 1970s, only eighty-nine breweries produced all U.S. domestic beer. *Chart courtesy of the Brewers Association, Boulder, Colorado.*

In response to light lager dominating the market, Americans longing to have greater diversity in beer styles and more flavorful options became homebrewers. In many cases the activity was illegal, even if the homebrewer was not selling the beer. The Twenty-first Amendment addressed home winemaking but did not mention homebrewing, leaving a broad gray area for legal interpretation. Individual states had varying legal holdovers from Prohibition. This didn't stop anyone from brewing at home. The general public was tired of Prohibition, and law enforcement in most areas had better things to do than chase brewing violations.

In October 1978, President Jimmy Carter signed House Resolution 1337, which contained an amendment that exempted taxes on beer brewed at home for personal or family use (in other words, homebrewing became federally legal). However, the Twenty-first Amendment still allowed states to self-regulate the trade of alcohol, and many held out for a long time. Homebrewing was not legal in Georgia until 1995, seventeen years after a Georgia-born-and-raised president signed in the federal exemption. It finally became legal to homebrew in Alabama and Mississippi in mid-2013. But it was a hard-won victory, with retail homebrewing equipment having been confiscated in Alabama as recently as 2012.

Underground homebrewing, now legally released from the shadows, sowed the seeds for what would become the modern craft beer movement. The father of American homebrewing, Charles "Charlie" Papazian, would publish the quintessential homebrewing guide *The Complete Joy of Home Brewing*. This guide described for the beer consumer the basic technical skills to produce his or her own beer. Not just any beer, but varying beer styles—including some that were nearly forgotten to history.

Microbrewery pioneers, such as Jack McAliffe of New Albion Brewing Company (1976), "Fritz" Maytag III of Anchor Brewing Company (1971), Ken Grossman of Sierra Nevada Brewing Company (1980) and John Maier of Alaskan Brewing Company and later Rogue Ales (1987, 1989), launched the American craft beer revolution using knowledge gained in homebrewing.

Early beer writers and journalists, chiefly Fred Eckhardt of Portland, Oregon, and Michael Jackson of England, helped bring about the modern beer renaissance and, more significantly, laid the framework for the modern concept of beer styles. In history, local terminology defined the name for a beer: the monk's beer, that white beer, so-and-so's red beer, the porter's beer, the strong ale and so on. The modern beer-style system categorizes beer by history, origin and custom, production method and/ or recipe, ingredients, color, flavor factors and strength or other unique factors. Michael Jackson's 1977 book, *The World Guide to Beer*, is considered a lasting reference on this concept.

The growth of craft beer in the western United States and the renewed interest in varying beer styles were influential in the Atlanta beer resurgence. Members of the Covert Hops Society, an early Atlanta homebrewing club, were instrumental in getting Georgia alcohol laws relaxed in the 1990s, leading to legalized homebrewing and brewpub operation. Early contract breweries (Friends Brewing Company and Georgia Brewing Company), early small Georgia breweries (Helenboch in Helen and Blind Man Ales of Athens) and early brewpubs would set the stage for the return of craft beer in Georgia.

CHAPTER SIX

BEER FROM AROUND HERE

Born and Brewed in Atlanta.
—*Dogwood Brewing Company's early slogan*

W hat's the darkest beer you got?"
This was the common question from a beer geek in the late 1980s and early 1990s in the Southeast after bellying up to the bar. Words like microbrew, craft beer and even local brewery (let alone the hundreds of names of various beer styles) were not commonly known yet. It was all about what is "darkest"—that is, what had the most malt content. If you were not a homebrewer or did not have access to homebrewed beer, you were lucky to find a beer that had any overt malt or hop flavor. An American pale ale of today would have been considered dark. An unfiltered beer would have been thought to be spoiled or contaminated in some way.

Into this uninformed market entered four Atlanta breweries. Words like crazy and reckless are common when describing the attempts to open a local brewery and produce a beer that did not taste exactly like what the mega-breweries were producing. The distributors were wary of carrying the products, and the retailers didn't want to give up valuable tap and shelf space, which equate to immediate profits. Only a very brave (or slightly off-kilter) person would launch a brewery in such a market.

Marthasville Brewing Company

Marthasville Brewing Company was Atlanta's first contemporary microbrewery, opening and beginning production in 1993. Marthasville was founded by Michael Gerard (sales and marketing), George Lamb (operations manager) and Doug Hubbard (brewer). Martha's Pale Ale, an unfiltered pale ale, was its first offering, but Sweet Georgia Brown Ale was one of the most popular beers it produced, followed closely by its Southern Ale.

Marthasville held the unenviable position of being the first contemporary craft brewery in the Atlanta area, not to mention one of the first in the Southeast. Resources that are commonplace for microbreweries now were harder to find in the Southeast during the early 1990s. By scrounging for brewing equipment from dairy supplies (at a fraction of the cost), bottling equipment from a soda manufacturing graveyard and other breweries' older filling and labeling equipment, Marthasville forged ahead.

At the time of Marthasville's formation, most types of brewing in Georgia had just become legal and operations were heavily scrutinized by local authorities. "It would have been easier to get a license to produce automatic weapons than get a license to brew beer," recalls George Lamb.

The crew remembers drinking in a bar one evening, worrying over the cost of spent grain disposal that would cost them $2,500 a month for a special sanitary-grade trash bin. In a twist of fate, an Alabama pig farmer was drinking next to them and overheard the conversation. Astounded that the Marthasville owners would have to pay a huge cost to dispose of what he saw as pig feed, he offered to take the grain off their hands at no cost.

However, not all unconventional operations went as planned. Using dairy equipment had its challenges. Dairy tanks are horizontal, whereas brewing tanks are typically vertical. This allows easy disposal of the dead yeast and sludge at the bottom of the beer tanks. This was not so with horizontal tanks; a brewer has to enter the tanks and "muck" the spent yeast and residue. The unfortunate biological truth is much of the sludge is still composed of living yeast and producing CO_2. The first employee to enter the tank passed out from lack of oxygen. Afterward, mucking was a two-person operation—one to muck with a rope tied around his waist (while a fan was blowing the CO_2 out of the tank) and a second outside the tank holding the rope to haul him out if he passed out from lack of oxygen.

Marthasville also kegged into an older-style metal keg with a side wooden bung. These defunct Schlitz kegs were procured at wholesale in the Midwest, and a Marthasville label was strategically placed onto the filled kegs. With these older kegs (1940s–70s), likely Hoff-Stevens or Golden Gate kegs, the used saturated wooden bung had to be removed from the keg before sanitation and a new bung placed before refilling. They were hard to keep sanitized but easy to roll and looked cool. As a result, the metal kegs were often stolen for frat parties, college décor and the occasional beer collector.

The Marthasville lineup included Harvest Ale, Wagon Ale, Atlanta Gold (an ultra-light lager), Southern Ale, Southern Stout, Classic Ale and Martha's Terminator Ale. The brewery employed fifteen people and brewed five thousand barrels of beer in 1995. However, Marthasville was slightly ahead of the craft beer revolution in Atlanta. Its early start was compounded by financial risks, overextension of funds and high investor turnover leading to the brewery's ultimate demise in January 1997.

ATLANTA BREWING COMPANY
AKA RED BRICK BREWING

The Atlanta Brewing Company (ABC) was founded in 1993 by Greg Kelly, a former Guinness executive. The brewery, a converted old red brick warehouse, was originally located at 1219 Williams Street. Pine-sided vats and cold tanks from England were purchased, and the beer was produced using an older traditional style of brewing. The large red malt mill, still used by the brewery today, was found sitting abandoned in a meadow in England. After being shipped to Atlanta and undergoing extensive refurbishing, the mill was put to work readying the grain for brewing.

The name "Red Brick Ale" was given to the brewery's flagship beer. One suggested (but ultimately discarded) name for the first beer was "John J. Bips Ale," named after a long-term brew master at the historic Atlantic Brewery. Atlanta Brewing Company produced four thousand barrels of beer in 1994, its first year of production. To maintain this initial demand, the brewery brewed once or twice a day. Peachtree Pale Ale, Spring Brew, Summer Brew, Winter Brew, Golden Lager and Kelly's Light were added to the beer lineup in the early years.

Among the various collaboration projects of 1996, ABC partnered with a Belgian brewery to release a Belgian beer called "Malone's." The beer was produced in Belgium, pasteurized and then shipped in a large tanker truck. Upon arriving in the United States via ship, the tanker was driven to the brewery, and the beer was pumped into a holding tank. This imported beer

The original location of the Atlanta Brewing Company, 1219 Williams Street Northwest. This site would be demolished for a highway expansion project. *Photo by and courtesy of Florian Vollmer.*

was then carbonized and bottled for distribution. Surprisingly, two out of three batches survived this bizarre trip in drinkable condition.

In 1997, Atlanta Brewing Company brewed Old Testy Coffee Stout for the Sandy Springs Taco Mac location. But the brewery's most notable and lasting partnership is with the Vortex Bar and Grill, a two-location Atlanta institution. The beer's labeling features the Vortex's iconic "Laughing Skull." The original full-bodied Bohemian pilsner was dry-hopped and lagered for sixty days before shipping out to thirsty fans.

Although moderate in success, the brewery was having hard times as it rolled into the twenty-first century. Portions of the brewery's equipment were temperamental and deemed challenging by the staff using them; quality control and management issues were also reported. Due to these inconsistencies, the distributor began to lose faith in the brewery's survival. Sales began to drop off, and the future of the brewery seemed uncertain. Adding to this frustration, the Georgia Department of Transportation's planned road expansion project doomed the brewery's Williams Street location.

In 2005, Robert "Bob" Budd was brought on by the brewery's investors as a management consultant to help with ABC's production issues. Once on board, Budd helped negotiate a settlement with the Georgia Department of Transportation. In 2006, Greg Kelly left Atlanta Brewing Company and Budd stepped up as president of the brewery. The remaining staff—including brew master Dave McClure—moved operations to a new location at 2323 Defoor Hills Road.

In 2009, ABC and McClure released the newest version of Laughing Skull, changing it from a Bohemian pilsner to an amber ale. By 2012, it had become the brewery's bestselling beer, while Red Brick Brown, Blonde, Pale Ale and Porter were still the four standard offerings of the brewery. Within the same year, the brewery began producing limited release beers under the Brick Mason series in twenty-two-ounce "bomber" bottles and special four-packs.

In 2010, the brewery changed its public name to Red Brick Brewing Company (via a DBA, or doing business as operating name). Brewery legend has it that the name for Red Brick Brown, and the new name for the brewery, originated from a historic speech. The speech was supposedly given by the Atlanta mayor shortly after General Sherman torched the city. He stated that the city would "rise from the ashes like a Phoenix" and be rebuilt "one red brick at a time."

As of mid-2013, the brewing crew includes head brewer Garett Lockhart backed by Steve Anderson, assistant brewer, and Nick Fowler, production engineer. Red Brick Brewing currently distributes its beer to eight states across the Southeast and produces 2.8 million bottles of beer per year (equivalent to ten thousand barrels). The company's annual sales growth is 40 to 45 percent. The current top seller is Hoplanta, a mild and balanced IPA.

Never in the company's history has morale been as high as it is today. Every member of the staff promotes the brewery's products and Atlanta craft beer as a lifestyle. After tweaks to the recipes and testing, the beers are better than ever. Award-winning beers are being produced by Red Brick Brewing, and a considerable barrel-aging program has produced impressive results, including the Anniversary Ale series aged in various bourbon barrels.

Bob Budd is a spirited advocate of Atlanta beer and craft beer in the Southeast. He has an understanding of the role of beer in the South's past and a sense of its potential future. For example, he wants to tap into the trend of "beercations." Much like wine tours of California, people from all over the country are now traveling to cities specifically for the local beers, brewery tours, bars and special beer events. Bob wants the Georgia craft beer industry to capitalize on this trend, and it seems that Red Brick Brewing—Atlanta's oldest contemporary brewery—will be there to help shape the future.

Dogwood Brewing Company

Traipsing across Europe with a copy of Michael Jackson's *Pocket Guide to Beer*, Crawford Moran, the man behind Dogwood Brewing Company, learned to appreciate well-crafted beer. Upon his return to the southern United States, it was apparent that the local beer culture lacked the flavorful beers he had come to love in Europe, as well as a proper place to enjoy them. While attending the University of Georgia, Moran was a regular at an Athens bar called the Globe. For operating in a college town, the Globe was unusual for its absence of televisions, a moderate selection of craft and import beers and a quiet-natured intellectual crowd. What was lacking, though, was a way to enjoy a wider array of fresh European-style beers.

A telephone call with a friend turned him on to homebrewing, and he began to experiment with re-creating beers he had found intriguing during his trip. In 1995, he attended the Siebel Institute, one of a handful of brewing schools in the United States. Bolstered with his early amateur brewing successes and constant urging from a friend that he should brew professionally, Moran began to seek financial backing and potential investors. During Dogwood's planning and financing stages, both Marthasville and Atlanta Brewing Company opened in Atlanta. He persevered, and at the ripe old age of twenty-eight, the Atlanta native got his brewery. Named after a southern tree with a lot of regional and cultural history in Atlanta, the Dogwood Brewing Company opened for business in 1996.

Drawing from his earlier homebrew recipes, Moran converted them to commercial brewing scale and added recipes developed in-house in collaboration with head brewer Matt Speece (a veteran of Rockies Brewing Company and Oasis Brewing) and assistant brewer Tony Camblor. Within a year, the budding brewery was winning awards for its flavorful beers. Dogwood produced three beers year-round: Pale Ale, Wheat Beer and Stout. During its lifespan, Dogwood Brewing produced between thirteen and fifteen distinctly different beers. Dogwood Pale Ale was ultimately the brewery's bestselling beer. All Dogwood beers were available in bottles and kegs, with the bottling line making up approximately 65 percent of the brewery's output of three thousand barrels (ninety-three thousand gallons) in 2000.

Dogwood's brewers were sticklers for using traditional ingredients and processes: using whole-leaf hops, maintaining in-house yeast strains, finding malt true to the history of a beer style and naturally carbonating their beers. Dogwood's pilsner was lagered for at least two months, a slow process more in tune with earlier brewing processes. Each year, the Dogwood Brewing Company produced a unique, limited-edition winter ale. The 1999 winter seasonal offering was a Belgian brown ale, developed using Chimay yeast. It was voted best local brew for that year by *Atlanta Constitution* writer Michael Skube.

In 2004, Dogwood released the first of its BrewMaster's series. The first beer in the series was Youngblood's Imperial Porter. Unfortunately, it would be the first and last. Dogwood Brewing closed its doors in 2004. The Brick Store Pub in Decatur kicked off the "wake" for the closed company in September 2004 by hosting the first of several events to drain the last kegs of Dogwood beer.

Of Atlanta's early contemporary breweries, Dogwood Brewing Company stands out as one of the most adventurous. In a time when pale ale was considered "dark beer," Dogwood was producing a multiple award–winning but commercially failing stout. Its brewers were also aging beer in bourbon barrels for special events long before the current trend, and the memories live on for many Atlanta beer geeks.

SweetWater Brewing Company

The story of SweetWater Brewing Company begins with University of Colorado roommates Freddy Bensch and Kevin McNerney working at a local brewery washing kegs. The craft beer industry was rapidly growing in the western United States, and Colorado was beer territory. This environment allowed the friends to develop a taste for craft beer and the industry that made it. After earning their respective degrees, Bensch and McNerney trained with the American Brewers' Guild. As brewers, they produced beer for various West Coast breweries, including Avery Brewing Company, Mammoth Brewing Company and Rockies Brewing Company.

The Olympic Games brought Bensch to Atlanta, and he came to the painful realization that the Southeast seriously lacked beer diversity. His longing for craft beer led to the reunion of the two brewers and the plan to build a local brewery that produced the hoppier beer styles with which they had become familiar on the West Coast. Add Matt Patterson, sales director in the venture (who uprooted himself and his then-pregnant wife to move to Atlanta from a job at the Breckenridge Brewery in Denver), and you have the beginnings of Atlanta's most-recognized brewery. "We threw our entire lives into the brewery, living and breathing the culture of a couple of young guys in a town with a young population. People could relate to what we were trying to do, and I think people like the underdog," recounts Bensch.

On February 17, 1997, SweetWater Brewing Company opened its doors at 900 Wendell Court, just off Fulton Industrial Boulevard on Atlanta's west side. The brewery was named after Sweetwater Creek, located roughly fifteen miles west of Downtown Atlanta. Bensch was kayaking the creek during the construction of the original brewery when the inspiration for the name struck. Despite the brewery being in a rough section of town, brewery tastings were a success, and its early canned food drives were "packed to the gills." SweetWater Extra Special Bitter (ESB) and Blue (a blueberry pale ale) were their original offerings, followed by 420 Pale Ale, which became the company's flagship beer.

The early days of SweetWater were a Bohemian dream. Beer flowed, the team members formed a family of their own, worn and comfy couches adorned the office space and guests were greeted by the brewery's mascot dog. The brewery once had a fifty-barrel (1,550-gallon) tank of beer spring a leak. Hard as the staff tried, it could not stop the leak, and there was no way to move the beer. Bensch recalls, "We just had to drink it! It was the never-ending beer fountain of 420. The dog was on it; everyone was on it. It was either a long, slow death or a big party, depending [on] how you look at it."

Friends, family and volunteers stopped by to set up boxes and pack bottles. The biggest volunteer turnout was for the brewery's Festive Ale bottling day. Festive Ale—a winter warmer–style ale with spices, available on a limited-release basis in the winter months—gained an almost cult following in its early days. In contrast, Georgia Brown (a full-bodied American brown ale) might be one of SweetWater's most underappreciated beers. Bensch attributes this to its lower production distributed across a smaller area.

In 2003, the quickly growing brewery had reached capacity. Its supply could never quite meet its demand. The company purchased additional equipment from a defunct California brewery and opened a new space on Ottley Drive, closer to Downtown Atlanta. Sporting a new higher-capacity brewery and a world-class tasting room, SweetWater Brewing Company continued to rise in the ranks of America's craft beer scene. The addition of the Dank Tank line—a unique high-gravity beer series with limited release—allowed the brewers to be creative and innovative while giving consumers a chance to try new flavors.

Again in 2012, the brewery expanded in size and production. In the same year, SweetWater ranked thirty-third among the top fifty U.S. breweries and twenty-fourth among the top fifty craft brewing companies based on sales volume. Clearly positioned as a regional brewery, SweetWater beers are distributed to Georgia, Alabama, Tennessee, Florida, North Carolina and South Carolina. By the end of 2013, it will produce roughly 142,000 barrels of beer (4,402,000 gallons) and extend distribution into Kentucky, Louisiana and Virginia.

However, Bensch feels that SweetWater is still a local brewery, firmly rooted in the community and faithful to the original core values on which it was founded. The company has expanded its community outreach, spreading awareness to support efforts such as clean water conservation through its "Save the Waters" campaign and giving back to community kids through its Camp Twin Lakes partnership.

THE CITY EMBRACES LOCAL BREWS

Atlanta taverns, bars and pubs that kept beer flowing during the years of mega-brewery mass production slowly became accustomed to local brews and, tap by tap, began to augment their selections with local beers. Bars such as Neighbor's, Manuel's Tavern, Atkins Park, the Vortex, Taco Mac, Summit's Wayside Tavern and many other venues in the Atlanta neighborhoods of Little Five Points, Virginia-Highland and Downtown Atlanta helped pave the way for craft beer's return to the metropolitan area.

By the mid- to late 1990s, Atlanta once again had local breweries. Visitors who asked about local beer were often shocked to hear a bartender give the affirmative, "Yes," and point them to a tap, maybe two, bearing local brand names. The names would change, but the first shots in the southern craft beer revolution had been fired.

Name	Location	Estimated Years of Operation	Notes
			Greater Atlanta's Post-Prohibition Brewery Graveyard
Atlantic Company	Atlanta, Georgia	1933–1955	Only post-Prohibition years are noted. Brewery spanned nearly one hundred years under various names.
Black Bear Brewing Company	Atlanta, Georgia	1996–2001	Contract brewery and part of a larger contract chain.
Carling Brewing Company	Atlanta, Georgia	1958–1972	Canada-based chain; produced Black Label, Red Cap Ale, Heidelberg Light Pilsener and Calgary Malt Liquor in the Atlanta brewery.
Dave & Mark Brewing Company	Atlanta, Georgia	1994–?	Contract brewery and maker of Black Sheep Lager.
Dogwood Brewing Company	Atlanta, Georgia	1996–2004	One of Atlanta's first contemporary craft breweries.
Friends Brewing Company	Helen and Atlanta, Georgia	1989–1998	Contract brewery and maker of Helenboch beers and Georgia Peach Wheat. On-site brewery in Helen operated 1990–1991.
Georgia Brewing Company	Atlanta, Georgia	1989–1992	Contract brewery and maker of Wild Boar Beers.

	GREATER ATLANTA'S POST-PROHIBITION BREWERY GRAVEYARD		
Name	Location	Estimated Years of Operation	Notes
Liberty Brewing	Atlanta, Georgia	1933–1933	Never produced. Company filed for permit at end of Prohibition. Denied due to ongoing state prohibition.
Marthasville Brewing Company	Atlanta, Georgia	1993–1997	One of Atlanta's first contemporary craft breweries.
South Atlanta Brewing	Atlanta, Georgia	1933–1933	Never produced. Company filed for permit at end of Prohibition. Denied due to ongoing state prohibition.
Southern States Brewing	Atlanta, Georgia	1933–1933	Never produced. Company filed for permit at end of Prohibition. Denied due to ongoing state prohibition.
Stone Mountain Brewers, Incorporated	Stone Mountain, Georgia	1992–?	No data available.
Zuma Brewing Company	Atlanta, Georgia	2004–2007	Makers of Cancun Beers.

CHAPTER SEVEN

BEER MEETS CUISINE

A Tale of the Brewpub

*Paying that little bit of attention, both to your food and to your beer, is the
difference between having an "OK" culinary life and having one filled with
boundless riches of flavor.*
—Garrett Oliver
The Brewmaster's Table, 2005

B eer and food have a long partnership, especially since beer was
considered a food item through most of its history. In the last fifty years
in the United States, classic beer pairings have been with pizza, bratwurst,
pretzels and hot dogs. The modern brewpub helped elevate this relationship
to matches of complex beer styles with more sophisticated food.

A brewpub is a pub or restaurant that produces its beer on the
premises, primarily for consumption on site. This concept is not new; in
fact, it is thousands of years old, originating in Europe. The brewpub as
a contemporary concept—higher-quality food and beer prepared in-house
by a chef and a brewer—started in the Northwest United States in the
early 1980s. It quickly became the laboratory for thoughtful pairings of the
aforementioned higher-quality food and beer.

In 1995, Georgia House of Representatives Bill 374 (HB 374) allowed eating
establishments to produce a limited quantity of beer in house for on-site draft
consumption. The brewpub had finally arrived in Georgia, conveniently in time
for the 1996 Olympics. Atlanta's brewpubs quickly became beloved institutions
and set the tempo for the city's modern beer and food scene.

MAX LAGER'S WOOD-FIRED GRILL & BREWERY

Located in the heart of Downtown Atlanta on Peachtree Street, Max Lager's opened in March 1998 and is the oldest continually owned and operated brewpub in Atlanta. John "JR" Roberts, an award-winning brewer, has held his position at Max Lager's since the start. Brewing a wide variety of beer styles, JR likes to keep his beers "sessionable" (easily drinkable for repeat sampling). His brewing caught the attention of the late, great international beer guru Michael Jackson, who visited Max Lager's three times. Shortly after producing his first batch of beer at Max Lager's, Roberts found himself sitting down in the brewpub with Ken Grossman, pioneer American craft brewer and founder of Sierra Nevada, and Michael Jackson. Excited but nervous, he received positive feedback from the beer legends.

Although not one to have a favorite among his many brews, Roberts is especially proud of his Imperial Mocha Oatmeal Stout (IMOS) and Hopsplosion!!!, an intensely hoppy IPA. In the summer of 2012, Max Lager's and Jailhouse Brewing Company produced a collaboration brew

John "JR" Roberts, brew master at Max Lager's Wood-Fired Grill & Brewery Restaurant. *Ron Smith.*

called Partners in Crime. This Berliner weissbier, a sour wheat beer made with traditional yeast and a lactobacillus culture (bacteria), immediately sold out. Max Lager's also produces small casks for "Firkin Nights" and occasionally produces a barrel-aged beer. Roberts is experimenting with bottle conditioning some of the house brews in 750-milliliter bottles.

Producing not only world-class beers but also freshly crafted food, Max Lager's is a favorite with downtown workers and travelers. It also serves as a convenient brewpub for the attendees of DragonCon, the popular yearly science fiction and fantasy conference held in Atlanta.

PARK TAVERN

This brewpub originally opened in 1996 as the Mill Brewery, Eatery and Bakery, part of a larger franchise business. Former franchise CEO Paul Smith bought the Atlanta location, reopening it as Park Tavern in 2000. Named for its proximity to Piedmont Park, the tavern is located in a building originally used as horse stables for local racing enthusiasts. The brewpub is known for its one-dollar pints when it is raining, sushi specials, music and many community events.

John Stuart, brewer for the former Mill Brewery franchise and a veteran of setting up many of the Mill Breweries, retained his position making beer at Park Tavern for many years. Stuart produced beer for Park Tavern in a prescribed English-style lineup: pale ale, India pale ale, amber and porter. In 2007, Stuart moved on to become head brewer at Green Man Brewery, a mainstay in Asheville, North Carolina. Park Tavern continues to operate an in-house brewery. Its extensive menu offers both fresh-fish dishes and burgers that are ground and formed on the premises daily.

5 SEASONS BREWING COMPANY

Rising from the ashes of the fallen Phoenix Brewing Company, the original location of 5 Seasons Brewing Company (in the Prado shopping center) opened in Sandy Springs in 2000. Restaurateur Dennis Lange and chef David Larkworthy bailed out the popular but failing business, quickly turning it into a powerhouse on the Atlanta brewpub scene. The initial beer was produced by Glenn Sprouse, former brew master of Phoenix Brewing. Inspired by living in Germany during his earlier years, Sprouse produced a mainly German-style beer profile. His signature beer, Seven Sisters Münchner, continues as a staple at the original location. Brewing alongside Glenn Sprouse in the early days of 5 Seasons, Brian "Spike" Buckowski (later of Terrapin Beer Company fame) helped keep the beer flowing. The Prado location was designed after the brewpub concept popularized in the Northwest, with a lodge feel created by exposed wood, visible brewing equipment and high ceilings, yet having an overall intimate, cozy ambiance.

In 2007, 5 Seasons Brewing Company expanded north, opening a second location in Alpharetta, a thriving suburb of Atlanta (reusing the site of the former Buckhead Brewery and Grill). This expansion would mark the return of former Dogwood Brewing Company's Crawford Moran to the Atlanta beer scene. After Dogwood's closure, he was looking to open a beer and pizza place in West Atlanta and had selected a location at 1000 Marietta Street.

After consulting with his friend, David Larkworthy, Moran tabled the pizza and beer concept and joined "Chef Dave" in a dual location 5 Seasons deal. Moran began managing the brewing at 5 Seasons Northside, and his Marietta Street pizza-friendly location opened in 2009 as the company's third restaurant, dubbed 5 Seasons Westside.

Open, airy and modern, 5 Seasons Westside was a departure from the décor of the other locations. The facilities are also a departure from the standard brewpub brewery. Having the largest brewing floor space of any brewpub east of the Rocky Mountains, 5 Seasons Westside gives Moran the flexibility to experiment with his favorite recipes. Known for his prolific skill in brewing saisons (French

Crawford Moran, brew master at 5 Seasons Brewing and former owner and brew master of Dogwood Brewing Company. *Ron Smith.*

for season), a light soft Belgian farmhouse–style beer with a combination of spicy and fruity flavors, Moran has brewed, barreled and bottle conditioned some of the best examples of the style that we have ever had.

Chef Larkworthy and brew master Moran play off each other in a match of prowess that results in brewing and gastronomic excellence. 5 Seasons Westside often hosts beer and food pairing dinners and other special events that highlight this dynamic duo's creativeness.

In 2008, former SweetWater cofounder and brew master Kevin McNerney joined the 5 Seasons Brewing team, replacing Sprouse at 5 Seasons Prado. McNerney brought to the 5 Seasons team over fifteen years of experience in the craft beer industry. In 2002, while at SweetWater Brewing Company, he was awarded Small Brewing Company Brewmaster of the Year at the Great American Beer Festival. His Hopgasm IPA, designed to be drunk alone or with most of the house meals, was an immediate hit for 5 Seasons.

McNerney has been quoted as saying that he enjoys the freedom afforded by a brewpub, where he can experiment with different brews and be closer to the consumers for feedback on his creations. His role at 5 Seasons has given him more time to spend with his family. In 2012, he returned to SweetWater as a guest brewer to help produce its Fifteenth-Anniversary Ale.

A final 5 Seasons note: Crawford Moran eventually got his pizza place. Russ Yates and Moran paired up to open the Slice & Pint, a pizza brewpub in Emory Village, in the summer of 2013.

Gordon Biersch Brewery Restaurant

Known for its German- and Bohemian-style beers brewed to the specifications of the Reinheitsgebot (a 1516 German Purity Law) and its made-from-scratch food, Gordon Biersch belongs to CraftWorks Restaurants and Breweries. This large company owns around two hundred restaurants operating under fourteen brand names across the United States. Located on Peachtree Street, the Midtown Atlanta Gordon Biersch opened in 1999. A second location opened later in Buckhead in the former Rock Bottom Restaurant and Brewery site (also a CraftWorks brand).

The vast majority of the beers Gordon Biersch produces are lagers—various bock, schwarzbier and pilsner recipes. This is unusual for Atlanta brewpubs, which predominantly brew ales. Lager brewing takes

Eric Geralds, regional brewer of Gordon Biersch Brewery Restaurant. *Ron Smith.*

more time due to its longer clarifying and maturing process; Gordon Biersch has to carefully forecast the brewpub's demand months in advance.

Eric Geralds is the regional brewer for Gordon Biersch. He aids the individual brewers within his region, making sure they have everything they need to brew. Both Geralds and the company are focused on consistency and quality control. He recalls the company bringing in two professors of fermentation science to help the staff learn and customize the process for each type of brewing system in the region. The service teams are trained by Geralds and the local brewers to know the characteristics of every beer and be able to answer customers' questions.

Although a national chain, Gordon Biersch's corporate philosophy is that it's vitally important to be focused on the community in which each store operates. The brewers are often members of the local brewers' guilds and take part in local events. Geralds hopes that the Atlanta locations will soon be able to sell growlers, adding their beers to the city's take-home options.

Twain's Billiards and Tap

Opened in 1996, Twain's Billiards and Tap sported a host of pool tables, shuffleboards, air hockey tables and pinball machines, along with a full bar. It offered good food to accompany the libations and entertainment. Successful with their neighborhood bar, Twain's owners decided to go in a different direction in a 2006 renovation, converting to a brewpub. Twain's first brewer was Jordon Fleetwood, formerly of Dogwood Brewing Company. He produced Heaven for Climate golden ale, Hell for Society stout, Mad Happy pale ale and Sloppy Squirrel red ale. The latter was named for a squirrel that invaded the tank room during initial brewing of the red ale and had to be ousted with a water hose.

Fleetwood would remain the lead brewer until late 2011, when noted homebrewer David Stein took over brewing the Tropicalia India Pale Ale series, Henbragon brown ale and several collaborative efforts with other breweries. Stein has since moved on to found Creature Comforts Brewing in Athens, Georgia. Chase Medlin moved up to head brewer of Twain's in 2013, producing a wide range of intriguing beers including a tequila barrel-aged IPA.

Ben Halter, winner of the Twain's Darkside Homebrew Challenge, produced a marvelous stout inspired by a Puerto Rican holiday cocktail. This heady stout was conditioned with rum-soaked vanilla beans and cinnamon sticks. Called Coquito Stout, it was brewed in house and put on tap. Adventurous brewing paired with great food preparation seems to be the mantra at Twain's, much to the delight of patrons of this beloved establishment.

Wrecking Bar Brewpub

Located in the Victor Kriegshaber house, a twentieth-century Victorian-style structure, the Wrecking Bar Brewpub visually stands out at the corner of Moreland and Austin Avenues. A lengthy labor of love by its owners and operators, Bob and Kristine Sandage, the building has been renovated as a combination brewpub, restaurant and event space. The building is remembered by many Atlantans as an "old abandoned building in the wooded lot" or as housing Wrecking Bar Architectural Antiques (1970–2005).

The Wrecking Bar opened in June 2011, bringing the communities of Inman Park and Little Five Points their first brewpub in an elegant, award-winning reuse of the historic building. The brewpub is named after the former antique store, and the event space is named after the early home's original name: the Marianna.

Brewer Bob Sandage, along with assistant brewer Neal Engleman, cranks out a wide spectrum of beers through the brewpub's seven-barrel brewery. The gleaming copper and stainless steel system was sourced from a defunct

brewpub in Scottsdale, Arizona. The brewers try to maintain a wide range of beers on tap to please most palates: high and low ABV, hoppy, malty, dark and light beer styles.

The restaurant and bar area is in the lower level of the building. Loaded with gray stone and warm-toned wood, the space evokes the feeling of a rathskeller or European cellar pub. The Wrecking Bar won *Creative Loafing*'s Best Atlanta Brewpub award in 2011 and 2012. The future promises many more barrel-aging projects and craft beer collaborations, along with the brewpub's standard lineup.

Bob Sandage, owner and brew master, and Neal Engleman, assistant brew master, of the Wrecking Bar Brewpub. *Ron Smith.*

CHERRY STREET BREWING COOPERATIVE

Nick Tanner, mastermind behind the Cherry Street Brewing Cooperative, started homebrewing at the age of twenty-one after seeing a friend produce a unique blueberry porter using simple homebrew equipment. A creative person with experience as a restaurateur, he became an even more dedicated brewer when a local Colorado homebrew shop owner kept saying, "No" to his suggested brewing projects. After finishing college in Colorado, he moved back to Cumming, Georgia, to help his dad open and run Rick Tanner's in the Vickery Plaza. When the local business market started improving, they decided to become a brewpub. After the exhaustive and confusing licensing process, the Tanners finally got their approvals on December 12, 2012. Nick immediately fired up the three-barrel brewing system and began producing beer that would win them awards mere months later.

Brewing along with Jonny Bradley and Dan Reingold, fellow homebrewers and Cherry Street Co-op crew members, Nick Tanner keeps eleven or twelve

Nick Tanner and Jonny Bradley of the Cherry Street Brewing Cooperative. *Ron Smith.*

different beers on draft at neighboring Tanner's. Cherry Street has a sizable local following, and in return, it is community oriented. The brewpub holds monthly charity events, spotlighting local homebrew recipes with its "Friend of the Farmer" beer series (Nick Tanner's nickname is Farmer). They even run local contests to name the different beers.

On the last Tuesday of each month, Cherry Street hosts a Tuesday Beer Bash with a special-release beer, a live band and a pint glass raffle. Once a year, it produces a 12-12-12 beer with twelve different hops and twelve malt varieties, aged for twelve months in honor of its licensing approval on

December 12, 2012. The co-op has also been experimenting with recipes in hopes of providing a gluten-free beer for patrons.

The love of beer and sense of community show in the brewing at the Cherry Street Brewing Co-op. Alongside the food, atmosphere and friendly service at Tanner's, the brewery will no doubt serve locals for many years to come.

CHEERS TO THE BREWPUB

Brewpubs have added dimension to the contemporary Atlanta beer scene over the years. Often overlooked among breweries, a few of the city's brewpubs represent the earliest existing brew houses in the metro area. Many offer quick turnaround of small batch or specialty beers that a larger brewery simply cannot afford to make due to equipment constraints and production schedules. If the legal trend continues, brewpubs might soon offer limited growler sales, allowing some of these unique brews to be consumed at home.

Whitehall Tavern, painting by Wilbur G. Kurtz, 1942. *Photo courtesy of the Kenan Research Center at the Atlanta History Center.*

A variety of early Atlanta City Brewery and Atlanta Brewing and Ice Company beer bottles (1800s to early 1900s). *Bottles available for photographing courtesy of Ken Jones. Ron Smith.*

Left: Atlanta Brewing and Ice Company small beer barrel, circa late 1800s to early 1900s, with Atlantic collectibles shown in the background. *Breweriana available for photographing courtesy of Ken Jones. Ron Smith.*

Below: The Tabernacle in present day. Its pastor and founder gave a sermon criticizing local politicians for standing in the way of Georgia prohibition on the opening day of the building. Throughout most of its history, the Tabernacle would be a prohibition stronghold. Ironically, it has become an entertainment venue that serves alcohol. *Ron Smith.*

Great American Hop Ale was a carbonated Prohibition beverage made by the American Beverage Company of Atlanta from roughly 1902 to 1912. The beverage was made with steeped Bohemian hops and contained 0.46 percent ABV. Although containing a legal level of alcohol for most states, it was outlawed in Kansas due to the saccharin content. *From the personal collection of Ron Smith.*

Steinerbru label, Atlantic Company's Atlanta brewery. *From the personal collection of Ron Smith.*

A collection of Atlantic Brewery cone-top and flat-top beer cans, circa 1940s–1950s. *Cans available for photographing courtesy of Tad Mitchell and Six Feet Under Pub & Fish House. Ron Smith.*

The early 1900s wood bar—complete with brass rail—in current-day Manuel's Tavern. This classy bar exemplifies Atlanta's bygone saloon era. *Ron Smith.*

Marthasville Brewing Company Southern Ale label. *From the personal collection of Ron Smith.*

Red Brick beer flows at a weekly tasting event held at the brewery located at 2323 Defoor Hills Road Northwest. *Ron Smith.*

Hoplanta India pale ale bottles in the bottling line at the Atlanta Brewing Company (Red Brick Brewing). *Ron Smith.*

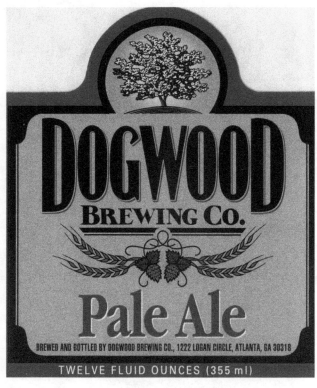

Dogwood Brewing Company Pale Ale label. *From the personal collection of Ron Smith.*

Dogwood Brewing Company bottles: BrewMaster's series Youngblood's Imperial Porter and the 2001 Limited Release Winter Ale. *Bottles available for photographing courtesy of Twain's Billiards & Tap. Ron Smith.*

Mural of the company's logo on the building of SweetWater Brewing Company. *Ron Smith.*

Signature trout tap handles proclaim the beer offerings at the SweetWater Brewing Company's tasting room. *Ron Smith.*

John Pierce and Brad Pass filling firkins (small one-quarter barrels) with Slammer Wheat at Jailhouse Brewing Company. *Ron Smith.*

The raised fist and loosened tie of a businessman is the logo of Monday Night Brewing. The "Great Wall of Ties" in its tasting room continues the theme, loosening the ties and stresses of Atlanta's hardworking folk. *Ron Smith.*

Marietta's Red Hare Brewing Company canning line fills cans of Gangway IPA. *Ron Smith.*

Above: The Belgian Beer Bar (upstairs) at the Brick Store Pub. *Ron Smith.*

Left: A portion of the Brick Store Pub's basement beer cellar. Note the old vault door. *Ron Smith.*

Above: Interior view of the Porter Beer Bar in Little Five Points. *Ron Smith.*

Right: The Midway Pub, located in East Atlanta Village, is a beer-centric bar that hosts a continuous stream of beer events, tap takeovers and specialty beers on tap. *Ron Smith.*

Der Biergarten, German restaurant and biergarten near Downtown Atlanta. *Ron Smith.*

Hop City's "Growler Town." *Ron Smith.*

Above: Display announcing Firkin Fridays at Crafty Draught in Cumming. *Ron Smith.*

Right: A tempting spread of signature appetizers and beer at Max Lager's Wood-Fired Grill & Brewery. *Ron Smith.*

Left: Twain's Billiards & Tap. Originally a restaurant, bar and gaming room, the establishment added a brewery in 2006, becoming Decatur's first brewpub. *Ron Smith*.

Below: A selection of beer from Twain's Billiards & Tap with flavors ranging from hoppy to malty to tangy from lactic acid. *Ron Smith*.

An overhead view of the brewery at the Wrecking Bar Brewpub. Sourced from a defunct brewpub in Scottsdale, Arizona, the equipment now produces beer for thirsty customers of this establishment on the border of Inman Park and Little Five Points. *Ron Smith.*

Burnt Hickory Brewery's Killdozer barleywine-style beer soaking in a bourbon barrel. *Ron Smith.*

Changing of the guard. The stylish taps at 5 Seasons Brewing's Prado location pour current brew master Kevin McNerney's signature red ale and Seven Sisters, a Munich-style beer originating from the location's first brew master, Glenn Sprouse. *Ron Smith.*

Exterior of 5 Seasons Restaurant and Brewery Westside, 1000 Marietta Street Northwest. *Ron Smith.*

GREATER ATLANTA'S BREWPUB GRAVEYARD

Name	Location	Estimated Years of Operation
Atlanta Beer Garten	(Buckhead) Atlanta, Georgia	1996–2002
Big Woody Brewing Company	Norcross, Georgia	2006–?
Buckhead Brewery & Grill	Alpharetta, Cumming, Peachtree City and Stockbridge, Georgia	1996–2007
Cherokee Brewing Company	(Buckhead) Atlanta, Georgia	1996–1997
Embers	Alpharetta, Georgia	2006–2007
Hops Restaurant, Bar & Brewery	Kennesaw, Douglasville and Duluth, Georgia	1997–2004
John Harvard's	(Buckhead) Atlanta and Roswell, Georgia	1995–2001
Mill Brewery, Eatery & Bakery	(Piedmont Park) Atlanta, Georgia	1996–2000
Pecker Head Brewery	Douglasville, Georgia	1998–2002
Percy's Fish House	(Buckhead) Atlanta, Georgia	1997–1999
Phoenix Brewing Company	Sandy Springs, Georgia	1996–1999
Rock Bottom	(Buckhead) Atlanta, Georgia	1997–2009
U.S. Border Brewery Cantina	Alpharetta, Georgia	1996–2003
Water Tower Brewing/U.S. Play	Kennesaw, Georgia	1997–2003

CHAPTER EIGHT

ATLANTA'S BREWERY EXPLOSION

The American microbrewery movement is the best that's happened to beer in 100 years.
—*Michael Jackson*
Atlanta Journal-Constitution, *November 22, 1999*

Atlanta breweries in the early 1990s faced a lukewarm craft beer market, reluctant distributors and strong competition within their ranks. By the following decade, the market had changed dramatically. Riding the wave of legal changes allowing higher ABV beer, better production rates and distribution and the ever-increasing reach of social media, newer breweries would find it easier to take root in native soil. Better communication led to improved relationships between the breweries and across levels of the three-tier system.

In 2010, a brewers' guild presence returned to Georgia as brewery after brewery announced plans to open in the Greater Atlanta area. By mid-2013, a total of seventeen breweries were either operating (local and contract), under construction or in the planning stage (see chart on pages 107–8). To the uninitiated, the number of new breweries sounds ludicrous. It's easy to think of rows of large factory-like breweries with semitrucks receiving loads of freshly brewed beer. In reality, the new breweries vary greatly in size and output—from a self-proclaimed nanobrewery in an office complex to Atlanta's second-largest brewery. The seven breweries reviewed represent this spectrum.

Burnt Hickory Brewery

A self-labeled nanobrewery, Burnt Hickory Brewery is located in a small industrial park in Kennesaw. Although quite small in size, the brewery's high-gravity (high ABV) beers are large in character. Owner and brewer Scott Hedeen is fueled by his loves of comics and music (punk, grunge and metal), with a kitschiness that shows in his brewing style and the brewery's labeling. The beers are named after local Civil War history, eclectic bands or whatever fusion strikes him at the moment.

Bursting onto the scene, Burnt Hickory first sold beer locally in 2012. Despite brewing with an amazingly small system, Hedeen has a fairly wide distribution and a solid following for his unique beers. Hedeen might brew in a frantic manner, albeit with diabolically clever results, but his tiny brewery packs in locals and beer geeks for the brewery's tastings and other social events. Beers such as Big Shanty Graham Cracker Stout, Didjits Lovesicle Blood Orange IPA, Cannon Dragger IPA and Ezekiel's Wheel Pale Ale suggest the beginning of a big, bold lineup of beers.

Eventide Brewing

Evenings on the front porch, fireflies in the air and stories being told by the elder family members gathered there. From the name (meaning evening or dusk) to their respectful approach to business and people, the folks at Eventide Brewing are focused on keeping things simple and true to tradition. Their motto reflects this concept: "Great doesn't have to be complicated." We assert that a good start for making the world a less complicated place is by crafting straightforward beer.

Moving into a space near Grant Park, Eventide plans to brew its draft beer on site and initially contract out the bottled beer until both can be produced locally. The initial brew is planned to be a kölsch-style beer, an ale originating from Cologne (Köln), Germany, that is warm fermented and then cold lagered to clarify and brighten the taste. Delicate on the palate, kölsch beers are popular with a wide variety of beer drinkers. An American pale ale and a stout designed to be nitrogen poured are also planned. The Eventide Crew—Nathan Cowan, Haley Cowan, Geoffrey Williams and Mathew Sweezey—target to begin production in fall 2013.

JAILHOUSE BREWING COMPANY

When Glenn Golden purchased space to open a brewery in his hometown of Hampton, he had no idea he'd purchased the old city jailhouse. Remains of bars, cells and other punitive paraphernalia were evidence of the building's history. Those discoveries ultimately inspired the branding of his venture as Jailhouse Brewing Company. The brewery began to take shape in 2008, as Golden's team procured brewing equipment that had been used at the former Buckhead Brewery and Grill in Stockbridge, Georgia.

Over the next year, the brewers created test batches and expanded their recipes to fit full-sized equipment. By the fall of 2009, Jailhouse was producing beer for the citizens of Henry County. Word spread quickly through social media, and production picked up to meet demand. As with most small breweries, Jailhouse had a difficult time competing for tap space in the local market. Having a solid relationship with a distributor focused on craft beer made all the difference.

Jailhouse has four year-round brews: Slammer Wheat, an American wheat ale; Misdemeanor Ale, a malty American red ale; Mugshot IPA; and Breakout Stout, a bold American stout. The brewery also produces seasonal beers for its Trustees series and offers one-time beers in the Solitary Confinement series. Golden encourages employees to run test batches on the brewery's small brewing setup for the employee specialty brew line called the Inmate Hooch series.

Currently, Jailhouse Brewing products are carried at craft beer locations all over Greater Atlanta. The brewing team is determined to maintain a quality product line and name brand. Their Deep, Deep, Deep, Deep Undercover India Pale Ale (aka 4D IPA) is a testament to their ability to explore brewing with various hops and malt.

Being a fan of small-scale organic farming and traditional European farmhouse ales, Golden originally conceived of opening a working resort farm and brewhouse in middle Tennessee. That dream ultimately gave way to the more practical creation of a local, high-quality brewery. However, his love for the farmhouse style of brewing has led to plans for producing Reprieve, the brewery's French-style saison, on an ongoing basis. It will be made using a traditional continuous brewing process. A portion of the fermenting beer will be added to the next batch, allowing the yeast to grow and evolve. Theoretically, the flavor profile of the continuously fermented beer should change as well. Jailhouse focuses on modifying ingredients while staying true to traditional brewing methods.

The Jailhouse Brewing Company crew, *from left to right*: Glenn Golden, owner and head brewer; John Pierce; Chris Broome; and Brad Pass. *Ron Smith.*

More barrel aging and further collaboration brewing are in the future plans of Jailhouse Brewing Company. Due to limited floor space, the brewery will eventually need to build out the existing space or expand to a new location. The team plans to always keep the original jailhouse for which the brewery is named.

MONDAY NIGHT BREWING

From the unlikely origins of Monday night Bible study, a group of acquaintances formed a bond over Monday night brewing sessions. Add promotion by social media, and Monday Night Brewing became a regular Atlanta social event (one that we regret never having attended). Over time, the Monday Night crew perfected a few beers and began planning to break into the industry. Developing their Eye Patch India pale ale, a sessionable but distinct IPA, was a five-year process. While refining recipes, the Monday Night team interviewed brewers, distributors and retailers for its blog. Such efforts added to its knowledge of Georgia alcohol laws and general marketing. This knowledge base would go a long way toward future beer sales. The group interviewed Kraig Torres of

Monday Night Brewing's (from left to right) Adam Bishop, head brewer and yeast whisperer; Joel Iverson, operations and taste-testing ninja; and Jonathan Baker, marketing guy and master of mind control. *Ron Smith.*

Hop City two weeks after he opened; Hop City would later become its top-selling account.

In the fall of 2011, Monday Night launched its initial beers: Drafty Kilt Scotch Ale, a Scottish-style ale with a balanced malty backbone and a hint of smoke, and Eye Patch Ale, a crisp and clean ale that lands in the area between hoppy American pale ale and a mild American-made IPA but with a distinct Magnum hop finish. For the first year, the beers were produced by Thomas Creek Brewery in Greenville, South Carolina, under the Monday Night brewers' supervision. Success of the beers was immediate. Monday Night Brewing was already known in the market due to its strategic use of social media and coordination with brewers, distributors and retailers.

By 2013, the new local brewery was ready for production. Brewing with a thirty-barrel system in a well-thought-out floor plan, it has room to grow. Monday Night's third offering, Fu ManBrew, is a Belgian-style white beer that is spicy and smooth with a hint of ginger. From the original formulation as a German hefeweisen, the brewers switched to Belgian yeast and used ginger instead of orange peel and coriander. This refreshing and snappy white beer took a bronze medal at the U.S. Beer Open. Recently released to market is Blind Pirate IPA, an 8.2 percent ABV double IPA.

Monday Night has a barrel-aging room that mimics the temperatures of changing seasons. They have plans to age Drafty Kilt Scotch Ale in bourbon barrels and perhaps do the same with to-be-developed recipes. The brewing team is working to create balanced beers that pair well with food.

O'Dempsey's Beer to Die For

Frustrated by yearly Canadian brewer strikes during hockey season, Randy Dempsey began brewing beer in his Toronto, Ontario home in 1984. The first batch was less than spectacular. "We went from sober to hung over, with nothing in between," comments Randy on the brew. Undaunted, he went to work on the bottling line of Conner's Brewing in an attempt to get closer to the brewers. They showed him the benefits of all-grain brewing and shared tips and techniques.

For years after gaining guidance, Dempsey brewed with better results and received many compliments. He brushed off most comments as general praise—that is, until over twenty years later. His boss—a serious beer aficionado—drove from Orlando to Atlanta to homebrew with Dempsey and wondered why he wasn't brewing professionally.

In 2008, Dempsey and Gail Smith, partner in O'Dempsey's, began looking into the legal aspects of running a brewery. They found the licensing process to be a nightmare, and the local government system was not set up to license a contract brewery. They persevered, and the first batch of O'Dempsey's Big Red Ale was brewed as an in-house beer at 5 Seasons Brewing's Prado location in cooperation with brew master Kevin McNerney. It was received well by patrons and staff, boosting confidence in the product's viability. Big Red Ale was initially commercially brewed and bottled by the Atlanta Brewing Company but moved to Thomas Creek Brewery as the contracted producer.

Randy Dempsey continues to work on developing new beers that follow his philosophy of creating balanced beer that works well at the dinner table. He extensively brews and tests the beers before working with Thomas Creek brewers to produce commercial quantities. Other beers now added to the lineup are Inukshuk IPA, Your Black Heart Russian imperial stout and Cold One pale ale. Proving his many talents, Dempsey designs the packaging and marketing, including the unique font found on O'Dempsey's labels.

Gail Smith sees O'Dempsey products as introductory beers for people who are branching out to artisan beer. The beers' easy-drinking qualities make them approachable. Both Smith and Dempsey are astonished at the change in the average consumer's knowledge. The beer drinkers of today

Gail Smith and Randy Dempsey of O'Dempsey's Beer to Die For toast to their success. *Ron Smith.*

are like the wine drinkers of the 1990s; the beer community is educating consumers the way the wine industry did two decades ago.

Dempsey and Smith are involved in the local brewing community and are appreciative of all the people who work hard to ensure the product is high quality and delivered properly—not just their products but also all the local beers. The name and beer lineup are well recognized in the Atlanta area. Dempsey, sitting down at a bar while awaiting a business transaction, struck up a conversation with the guy sitting next to him. Recognizing the name O'Dempsey's, the fellow launched into a nonstop dialogue on how much he loved the beer, recounting all the finer points. The O'Dempsey's lover never knew he was talking to the owner/brewer. Nothing speaks more highly of your product than having your beer promoted back to you by a consumer.

RED HARE BREWING COMPANY

Roger Davis and Bobby Thomas had been successfully brewing in Davis's basement for nearly two years when they realized their hobby could become a business. The process of forming Red Hare took three years and involved locating a brewing space, sourcing equipment, negotiating financial backing and completing necessary licensing. The name Red Hare is a spin on red hair, due to head brewer Thomas's distinctive hair color.

Red Hare Brewing is located in Marietta within an eleven-thousand-square-foot former patio enclosure business space. The crew personally did the tasting room build-out, adding finishing touches moments before the doors opened. The first patron to belly up to the tasting bar left a pint glass ring and an elbow dimple in the still-curing epoxy.

In September 2011, Red Hare opened to the public for tastings and tours, building up the brand name and generating market excitement. Shortly after, its beer was available in kegs for sale in Atlanta, Decatur and Marietta. In August 2012, Red Hare made history by being the first Georgia craft brewery to produce beer in cans. Stacks on stacks of empty cans awaiting the filling line dominate one corner of the brewery.

The brewery's 20-barrel brewhouse with a 220-barrel fermenting capacity produces its flagship beer, Long Day lager, and mainstays, Watership Brown and Gangway IPA. Red Hare has the added distinction of being the only local craft production brewery to consistently produce a lager beer. The lagering (cold storage) process takes more equipment and has a longer production time than most ale brewing. Beyond its year-round beers, the brewery produces limited-release beers called the Rabbit's Reserve series. This line of specialty beers has included a chocolate porter, an imperial red IPA, a saison, a black IPA and an Oktoberfest-style beer.

Red Hare won medals at the 2011 Grayson Blues and Brews Festival in the People's Choice category—gold for Long Day lager and bronze for Gangway IPA. The brewery also placed first in the Specialty Beer category at the 2012 Atlanta Cask Ale Festival.

STRAWN BREWING COMPANY

Will Strawn and Doug Evans can be found most afternoons at the Strawn Brewing Company. Hardworking men at their day jobs, they work equally hard at the self-built and operated brewery located in Fairburn. Housed in a portion of a former ice factory, Strawn is Fairburn's first brewery and currently one of the newest additions to the Atlanta brewing scene. Similar to an earlier iconic Georgia brewery (Marthasville), Strawn is built around repurposed food-grade stainless steel equipment. Sourcing equipment from the dairy, soup and toothpaste manufacturing industries, Strawn Brewing

Doug Evans pitches yeast into the fermenter at Strawn Brewery, Fairburn. *Ron Smith.*

has managed to keep its opening costs low, preserving funds for future expansion.

Opened on Labor Day weekend 2012, the brewery began making beer to be distributed across Georgia. The current offerings include wheat ale, Scottish ale and amber ale. Will Strawn reports that the amber ale is the bestselling beer. Patagonia malt gives the amber a slightly smoky-caramel flavor that is balanced by the hops. A hint of banana is imparted by the yeast, giving the beer a unique and refreshing flavor for the style. Strawn Brewery maintains the average ABV between 4 and 6 percent to create easy-drinking session beers. Its soon-to-be-released India pale ale will be a break from this trend with a slightly higher ABV.

WELCOME HOME, BREWERS

In our cross-section, both the brewers and the beers they produce show unique characteristics. Other breweries in the planning stages or soon to come online are also distinct from one another. By offering a variety of approaches and flavor profiles, Atlanta's breweries are reaching new followers. At the same time, local breweries are finding a more welcoming home on local pub taps. In the next chapter, we will visit a few of the pubs where beer-loving residents and visitors can enjoy the fruits of Atlanta brewing—as well as a diverse world of brewing styles.

BREWERIES OF THE GREATER ATLANTA AREA (AS OF JUNE 2013)		
Name	Location	Status
Atlanta Brewing Company, aka Red Brick	Atlanta, Georgia	Operating since 1993.
BlueTarp Brewing Company	Decatur, Georgia	Operating since 2012.
Burnt Hickory Brewery	Kennesaw, Georgia	Operating since 2011.
Eventide Brewing	Atlanta, Georgia	Under construction.
Jailhouse Brewing Company	Hampton, Georgia	Operating since 2009.
Jekyll Brewing	Alpharetta, Georgia	Under construction.
Monday Night Brewing	Atlanta, Georgia	Contract brewed by Thomas Creek Brewery, South Carolina, 2011–2012. Operating locally since 2013.
Monkey Wrench Brewing Company	Snellville, Georgia	Planning phase.
O'Dempsey's	Atlanta, Georgia	Contract brewed by Thomas Creek Brewery, South Carolina. Plans in place for local brewery.
Orpheus Brewing	Atlanta, Georgia	Planning phase.
Red Hare Brewing Company	Marietta, Georgia	Operating since 2011.
Second Self Beer Company	Lilburn, Georgia	Planning phase.

Breweries of the Greater Atlanta Area (as of June 2013)		
Name	Location	Status
Strawn Brewing Company	Fairburn, Georgia	Operating since 2012.
SweetWater Brewing Company	Atlanta, Georgia	Operating since 1997.
Terminus Brewery	Atlanta, Georgia	Planning phase.
Three Taverns Brewery	Decatur, Georgia	Under construction.
Wild Heaven Craft Beers	Decatur, Georgia	Contract brewed by Thomas Creek Brewery, South Carolina. Plans in place for local brewery.

CHAPTER NINE

A BEER GEEK WALKS INTO AN ATLANTA BEER BAR

Beer is what you drink as you sit around to figure out it, *whatever* it *is.*
—*Dave Blanchard*
Coowner of Brick Store Pub
Love at the Pub, *2009*

B eer has evolved from a regionally produced food source to occupying an epicurean realm of its own. Local styles have been studied, reproduced outside their historic range and sought out by a growing crowd of beer connoisseurs. Although many beer-centered bars exist across the United States, a few stand out by focusing on every detail of the drink.

In the 1980s and early 1990s, bars were judged on the number of beers they had available. A bar was world class if it had hundreds or even thousands of beers. A good example of this trend was the defunct Brickskeller in Washington, D.C., with over one thousand beers on its menu. That trend has changed to a greater focus on the quality and handling of strategic offerings. A state-of-the-art beer bar now offers a well-planned selection, the proper glassware for each style (rinsed before the pour), a food menu that supports the beers, a knowledgeable staff that can speak about the individual brews and (sometimes) a carefully cellared choice of specialty beers.

What you typically do not find in these bars is sophistication for its own sake, frozen pint glasses, poor-quality food and an array of high-definition television screens. Not that watching the game on TV is a bad thing, but it can be done anywhere. What is special today is world-class beer paired with excellent food and enjoyed over conversation with friends and family.

These extraordinary pubs become sanctuaries for traveling beer geeks, to the point of being destinations unto themselves in a beer-centered vacation. Ask anyone in the Greater Atlanta area who loves craft beer what local bars best fit this model, and the typical answers are the Brick Store Pub and the Porter Beer Bar.

PURSUIT OF THE EXCEPTIONAL PUB

Brick Store Pub

From the moment it opened in 1997, the Brick Store Pub set a new standard in Georgia beer bars. Loosely based on the Globe in Athens, Georgia, the Brick Store Pub is the brainchild of Dave Blanchard, Mike Gallagher and Tom Moore. The trio shared a vision for a European-style pub serving quality beer and food in a relaxed atmosphere. Located on the city square in Decatur, the pub occupies the site of an old general store that sold, among its sundries, bricks.

The relaxed atmosphere equates to no televisions, neon signs or loud music. Exposed brick, organic wood, earthy paint, reused antique doors, sparse but cool Belgian breweriana, locally sourced artwork and subtle stained glass are de rigueur at the Brick Store. The wait staff is not composed of beer experts only, but most team members are. The servers flow through the crowd with practiced efficiency that is not accidental. Management and staff have a level of respect for each other seldom seen in the service industry, so much so that Mary Jane Mahan wrote of the atmosphere in her book *Love at the Pub*.

Once ordered, the beers arrive quickly and in the correct glassware—not just correct for the style, but in the exact glassware. Your obscure Belgian beer comes out in the brewery's glass, which was freshly rinsed at the rinse station. Your second beer, an Allagash, comes out in Allagash's signature glass, and so on. There must be thousands of glasses stored somewhere in the establishment. This European tradition, adopted by the Brick Store Pub, is becoming an Atlanta standard. Respect the beer while drinking as the brewer respected the beer while brewing. "It is the Brick Store's culture to keep up with the little things, and [the] staff is trained to know the little things," confirms Blanchard.

Below the Brick Store's first floor, beer silently ages, awaiting patrons with a taste for cellared beer. The cellar winds through old storerooms, a bank vault and other seldom-seen areas. The walls bear signatures of noted American brewers and celebrated guests, and the racks include many eye-popping rare beers. The concept of cellaring beer is still in its infancy in America; this cellar was inspired by the owners' 2005 trip to Belgium.

Upstairs is the Belgian beer bar, built in 2004 when the ABV limit was raised in Georgia to a blessed 14 percent. Looking more like a European downstairs than an American upstairs, the Belgian bar pays homage to the mastery of Belgian brewing. Sporting nooks and beer shelves, the upstairs bar has seen many groups, classes and beer tastings. Interestingly, the dedicated Belgian beer bar came about by accident. As

Top: Dave Blanchard, coowner and operator of the Brick Store Pub in Decatur. *Ron Smith.*

Right: The upstairs beer cellar and tasting room at Decatur's Brick Store Pub. *Ron Smith.*

the beer market surged with the new ABV limit, the Brick Store had to figure out how to manage its increased stock. While organizing one day, Blanchard serendipitously said, "Let's move all the Belgian beers upstairs" because the majority of the new stock was Belgian. The rest is history.

As you sit in the Brick Store looking out at the patrons—their glasses streaked with Belgian lace—you see a crowd exhibiting a mix of elements: loosened ties, kicked-off dress shoes, dreadlocks, tattoos, jeans, skirts and T-shirts. Sprinkled in the beautiful din of human conversation are accents with international and southern origins. The wait staff is similarly mixed. But appearances are not relevant—it's Decatur; we're here about beer, and little else matters. In case you are not yet persuaded to visit, perhaps we should mention that an international beer geek website has voted the Brick Store Pub the number two beer bar on planet Earth.

The Porter Beer Bar

Following the model of the Brick Store Pub in concept, but a gem to itself in flavor and style, the Porter Beer Bar opened in September 2008. The establishment is located in the heart of the Little Five Points community, a historic and colorful section of Atlanta. The Porter is owned by Molly Gunn and Nick Rutherford, veteran restaurateurs, beer geeks and residents of Little Five Points. Wanting to bring into the area a restaurant with a large beer selection, they were nevertheless concerned with craft and import beer sales. Frequenting their local Vortex Bar and Grill, Gunn and Rutherford were astounded at the sales volume of Delirium Tremens, a strong golden ale produced by the Huyghe Brewery in Belgium. "I knew Little Five Points was ready for great beer when I saw that," Gunn recalls.

Decorated with simple porter-themed old suitcases and antiques, the Porter exhibits casual earthy elegance and a feel of instant history. The building's narrow structure gives it a sense of antiquity, reminiscent of narrow 1800s saloons and old European pubs. Surprisingly, its narrow footprint comes from the space's original use as an alleyway. The doorway and steps located between the barroom and dining area were once the back exit of a music store.

The Porter has forty-four draft lines, two hand-pumped stations and more than eight hundred bottles. The eyes of the Gulden Draak (Golden Dragon) tap light up while pouring the namesake beer. One tap tower sports the Delirium beer's signature elephant, a nod to their initial decision to open the Porter. Gunn likes to keep one tap available to experiment with unique

The Porter Beer Bar's beer cellar and event space, featuring a locally sourced and handcrafted wood table, Little Five Points. *Ron Smith.*

styles, unusual brews or something fun. The Porter often has special tastings and might roll out a pin, a small 5.4 U.S. gallon cask, for these events.

In 2012, the Porter expanded into the space next door, increasing its storage space and building the beer cellar. Exposed brick walls are lined with wooden shelves filled with bottles of beer maintained at sixty-three to sixty-five degrees. Niches contain Belgian breweriana and custom art. Central to the cozy room is a handmade table created from locally reclaimed deodar cedar.

Beyond its local following, the Porter Beer Bar has received national notice—appearing in *Draft* magazine's America's 100 Best Beer Bars several times. Ratebeer.com named it the best beer bar in the United States in 2012 and 2013, and the Porter was a James Beard Foundation semifinalist for the country's most Outstanding Bar Program in 2012 and 2013. Reaching an even broader audience, the bar was featured on *Anthony Bourdain: The Layover* on the Travel Channel.

HONORING ATLANTA'S ECLECTIC BEER BARS

Although breweries are becoming more accessible to the Atlanta public through tours, tastings and events, most people become familiar with the modern range of beer at the local watering hole. Albeit, the local watering hole has to have a decent choice of beer styles. Atlanta has hundreds (possibly thousands) of bars, pubs, taverns, grills, restaurants and even coffee shops offering beer on the menu.

We have selected a few beer-centered establishments that provide a taste of what the Atlanta area has to offer. Irish, German, corporate and individual—these bars do not merely present an array of beer but also promote knowledge through tastings and events and host themed dinners showcasing their food and beer.

Augustine's

Located on Memorial Drive near the historic Oakland Cemetery, this pub is a hidden gem. It originally opened as the Standard, one of several gas-station-to-restaurant concept locations built by a local chain. Bought out and redesigned by one of the former owners, Augustine's (formerly Young Augustine's) changed its menu and shifted to a beer-centric establishment in 2010.

The large chalkboard proclaims what's pouring from the numerous taps. A moderate bottle list augments the taps, and the staff is very knowledgeable about the beer served. Lots of window area lets in light to the modern wooden bar and table space. If Jerome is behind the bar, classic heavy metal can be heard at low volume while you enjoy your choice of craft beer together with upscale pub grub.

Over the past three years, Augustine's has participated in many events and offered specialty brews for the events. Julian, the current "beer guy," keeps the selection fresh and wide ranging.

Der Biergarten

Atlanta is familiar with German social institutions, brewers and beer (as recounted in chapter two). In the fall of 2010, this historical influence was reintroduced with the opening of Der Biergarten, located in the Luckie-Marietta District of Downtown Atlanta. The concept was the creation of Legacy Restaurant Partners and coowner Wolfgang Hartert, an Atlanta resident and German native who imparted authenticity to the establishment. Der Biergarten offers a good selection of Deutschland's finest beers on draft poured in traditional half-liter and liter sizes.

The dining room features white tablecloths and a wooden bar with a German model train circling overhead on a suspended track. Patrons enjoying the wurst (sausages) and schnitzels are surrounded by Bavarian murals. The indoor/outdoor beer garden has traditional Munich family-style seating on A-frame wooden tables. Der Biergarten was listed in delish. com's 9 Cool Beer Gardens Around the U.S.A. Prost!

The Marlay House

Fish and chips plus Guinness, Bass and Boddingtons—yes, these items are on offer, but that does not tell the whole story. The pub is owned and operated by the Comer family, which hails from Rathfarnham near Dublin, Ireland. Like the owners, the Marlay House has evolved from its Irish roots into a unique hybrid that fits well with the customer base in Decatur. Colin Comer, coowner and beer lover, recalls having to catch up with the U.S. beer scene when he first arrived from Ireland.

The Comers not only caught up, but their pub has also exceeded the typical bar. The Marlay has twenty-three tap lines that rotate a varied offering

The Marlay House in Decatur. *Ron Smith.*

of beer. The owners are known to hold back special kegs for events at the establishment. In keeping with the bar's roots, the televisions allow patrons to catch up on European soccer and cycling while enjoying a pint. On the next visit, that person might enjoy an American beer and food pairing. The Marlay has hosted a homebrewing competition on its patio while tapping some unique beers.

The food has evolved along with the beer. The Marlay hired an executive chef and adjusted its menu, adding nontraditional and farm-to-table items. A few of our favorites are risotto balls; the turkey, cranberry and brie sandwich; and garlic-roasted summer squash. The Marlay House continues to thrive. Once a drinking spot, it's now a neighborhood institution with long-term customers, ex-pats and entire families sitting down for food and drink. Sláinte!

Midway Pub

Wanting to open a neighborhood bar with a solid beer list and above-par bar food while remaining relaxed in its approach, owners Jonathan McIntyre and Carsten Green opened Midway Pub in East Atlanta Village in 2008.

The pub has a long bar that frames the taps and glass-fronted coolers, acting as a beer shrine. The beer list is well planned and makes up part of its "Pharmacy" along with other libations. The beer events and thirty-two frequently rotating taps bring in a constant stream of beer aficionados. The main seating area has garage doors between the interior and patio, an adjacent game room and a dog-friendly atmosphere. For the sports minded, a manageable number of TVs project every known athletic event.

In 2013, Midway Pub made CNN Travel's Best American Sports Bar list. The owners credit their staff and the enthusiasm of local residents for the pub's success.

Six Feet Under Pub & Fish House

Tad and Nancy Mitchell opened the first Six Feet Under Pub & Fish House in 2003 on Memorial Drive, naming it after their neighbors in Oakland Cemetery who are "six feet under." Six Feet Under is a unique beer destination in Atlanta. It has a selection of beers on tap but offers much more than that.

Six Feet Under, now at two locations (the second is on Eleventh Street), houses what is possibly the largest public display of breweriana in the Southeast. The collection belongs to coowner Tad Mitchell, who started acquiring beer cans in the 1970s at the ripe old age of ten. The collection was housed in the attic of the Mitchells' house until it was moved to the restaurants. Although appearing chaotic to

Atlantic Brewery breweriana inside Six Feet Under Pub & Fish House, Grant Park community of Atlanta. *Ron Smith.*

the casual eye, the objects are in alphabetical order. The Memorial Drive–Grant Park location starts the alphabet with an impressive display of Atlantic Brewery memorabilia. Old prints and posters decorate the walls of the staircase leading up to the patio, which provides views of the Atlanta skyline.

If you check out the Atlantic Brewery collection at the Grant Park location, raise a pint to Egidius Fechter—the godfather of brewing in Atlanta—who is lying in repose across the street in historic Oakland Cemetery.

Taco Mac

The Taco Mac story begins in 1979 in the Virginia-Highland neighborhood with an unusually named buffalo wing and beer joint. During renovation of an old taco stand to start their business, the original owners were a bit strapped for cash. Instead of shorting critical equipment, they decided to leave the old name on the building.

Along the way, the Tappan Street Group acquired the Taco Mac enterprise. Currently sporting twenty-eight locations (many of them in and

The original Taco Mac location in Atlanta's Virginia-Highland community. *Ron Smith.*

around Atlanta), Taco Mac has become a beer and food empire. Offering a large variety of beer on draft and in bottles at each location was a daunting task. Enter Fred Crudder, Taco Mac's beverage director, who coordinates the relationships with distributors and brewers (not to mention supporting the occasional journalist, blogger and highly interested craft beer geek).

In the beginning, each location had a random selection of beers on tap. The beer offerings have been restructured to ensure that solid national, regional and local beers are represented. In most locations, the local beers are in a "top center" placement among the tap rows. Crudder has seen the market change over the years, and today, around 75 percent of Taco Mac's beer stock is of craft brands. Within the Atlanta-area stores, SweetWater has set company sales records, and the brewery's 420 Pale Ale is always on draft.

But it's not all work for Fred Crudder. He has attended Sierra Nevada beer camp twice, one of the very few people to attend more than one year. Each year at camp, he guest-brewed red IPAs (with experimental hops) that were later available at select locations. In 2013, he will help Taco Mac host its second Southern Brewers Challenge. The 2012 winner brewed his beer at Atlanta Brewing Company, and it was available on tap at Taco Mac. There's also a speakeasy named for Crudder (but we just might be mistaken).

Taco Mac strives to enrich the beer knowledge of its patrons. Customers might be inspired to try new brews, but Taco Mac's business model has influenced managers and owners of other Atlanta-area beer bars, such as Summits Wayside Tavern and the Square Pub, to further advance the Atlanta beer scene.

A Toast to Atlanta's Bar Evolution

The Atlanta beer bar has come a long way from the stereotypical joint with its smoky interior, neon signs, peanuts and pretzels in bowls on the bar, a beer tap or two and a jukebox. Not to say a good, hometown dive bar isn't awesome, but what is remarkable is Atlanta's move toward beer diversity, adding another dimension to its history.

With the current beer renaissance inspiring variety, craft beer drinkers seek broader options at most every bar and restaurant. Urban and suburban establishments alike typically feature at least two craft beers as well as locally produced brews. We would argue that with this trend, Atlanta is on its way to developing its own barley-inspired cultural identity.

CHAPTER TEN

BEER STORES, GROWLERS AND BEER FESTS—OH MY!

It [beer] *holds as high favor with the man of wealth in his palatial club as it does with the humble laborer, who rushes the growler with his spare pennies.*
—*The* Constitution, *1895*

The Greenbaum family held one of the first liquor licenses in the state of Georgia when they founded Green's liquor store on Ponce de Leon Avenue in 1937–38. A member of the same family opened the first Tower Beer, Wine & Spirits store in 1948. Both companies would grow into chains, with Green's located in two states. These businesses became prime beer locations in the Atlanta area. Green's in particular would eventually offer a specialized selection of import and craft American beer. However, it would be the twenty-first century before the Atlanta area would see a similar, new business entity that focused primarily on beer.

SPECIALTY BEER STORES

Hop City

As the first specialty beer store, Hop City became ground zero for take-home craft beer in Atlanta. Owner Kraig Torres frequented several package stores around the Atlanta area and thought he could do a better job of beer sales.

Hop City opened in April 2009, two weeks after its grand opening party that was held in a partially completed space in the newly built 5 Seasons Westside. The opening days were very successful; Kraig could barely stock the shelves before the beer was sold. He attributes this success to social media that kept customers aware of the store's opening date and the available beers.

Hop City was the first store to organize stock by the style of beer. The shop is designed to be user friendly for both diehards and people new to craft beer. At first small, the homebrewing section is the fastest-growing segment of Kraig's business, amounting to 15 percent of total sales for the first part of 2013. In early 2011, Hop City expanded to open Growler Town, a sixty-tap growler-filling station. Hop City was one of three businesses that helped gain clarification on a Georgia law, ensuring that beer growlers were a legal container for beer in the state.

In 2012, Hop City opened a second store in Birmingham, Alabama, just before homebrewing became legal in that state. Plans are in place to open a third store in Atlanta, and Kraig hopes to open a new store every few years.

Ale Yeah! Craft Beer Market

Eddie Holley was no stranger to running his own business, but the struggling economy made changes necessary. Teaming up with business partners Cisco Vila and Jay Edwards, he decided to open a beer store in the Atlanta area. He was certain he wanted to stay self-employed, and his interest in craft beer had grown over the prior decade. He recalls trying a Sam Adams at a company mixer. Holley, like many new craft beer enthusiasts, launched his love for microbrews from one of America's midsized breweries.

The partners' first attempt to open a store in East Atlanta Village failed due to tight local regulations originally designed to limit the total number of package stores and gas station retailers of alcohol in a given area. A new spot on Decatur's College Avenue bore fruit in 2010 when Ale Yeah! became Decatur's first specialty beer store. Shelves are filled with a selection of international beers arranged by style, primary ingredients or tasting notes (e.g., sour). Local homebrewers can find supplies to make their own beers without traversing metro Atlanta. The store also offers local specialty cheeses, charcuterie, jams and mustards. In 2011, taps were added on the back wall so that Ale Yeah! growlers could be filled.

Ale Yeah!, as did several of the new beer businesses, engaged the power of social media and free networking to spread awareness of the company and blow-

Russ Goggans of Ale Yeah! Craft Beer Market in Roswell fills a growler for a customer. *Ron Smith.*

by-blow beer availability to Atlanta's beer geeks. The Ale Yeah! employees are passionate about good beer, which shows in their talking with excitement about brews during short breaks between the steady stream of customers. Basing off the success of the Decatur location, a second store was opened in 2013 in Roswell, giving its residents a new venue for meeting their craft beer needs.

RETURN OF THE GROWLER

No, it is not the title of a classic horror movie; it's a beer business model sweeping the Greater Atlanta area. Getting beer "to go" in one large container doesn't sound like a miraculous event. However, getting draft beer to take home is a recent return to an old practice. As noted in chapter two, the growler was common in Atlanta beer history. A growler could be anything from a can, to a pail, to whatever else held draft beer to be consumed away from the point of sale. The concept was demonized during the temperance movement, and all consumable alcohol was illegal during Prohibition.

In the decades after Prohibition, a multitude of state and local open-container laws were in effect. And let's face it: an open pail is not a great way to carry beer, keep it carbonated or protect the taste. In the Midwest and Great Lakes regions (the big brewery states), growlers returned in a limited way as large glass containers, similar to lemonade or milk containers. However, growler use did not immediately return in the South. Take-home options were single or six-pack bottles and cans; draft beer was for on-site consumption only. This situation existed until the law was clarified (that is, what exactly constitutes an open container versus a closed container) and plastic shrink-wrap was employed to close the container. Acceptable practices and how a container qualifies as "closed" are often determined by local ordinance.

The modern growler is typically a sixty-four-ounce (half-gallon) or thirty-two-ounce (quarter-gallon) dark-pigmented glass jug with a screw-top or flip-top cap. Other sizes can be found, such as one-liter, European-style growlers. As a bonus, growlers are environmentally friendly. The containers are used multiple times, reducing the overall amount of glass needed to transport the beverage.

Once the laws were made clear, Atlanta again "rushed the growler"—in a big way. In an amazingly short period of time, more than fifty locations in and around Atlanta now have tap stations. These taps appear in a variety of establishments including small delis/markets, supermarkets and even gas stations. Dedicated growler shops make up ten to fifteen of these locations. Below, we highlight two of these neighborhood growler shops to give a glimpse of this trend.

Beer Growler Nation

Kelly and Constantine Mihalis paired beer and food while entertaining guests at their home. Constantine often picked up the beer in a growler once this format became available, and he became intrigued by the business. Convinced that this line of business could use a different approach, they opened Beer Growler Nation in Decatur's Oak Grove Shopping Center. Their use of reclaimed items (old doors and cabinets) gives the growler shop a bit of rustic chic, but it is their warm and friendly service that most stands out. The couple views the shop as an extension of having people over to their home. They excel in providing a great experience to both beer novices and experts. The crew is quite knowledgeable; most have prior experience in the industry.

Beer Growler Nation is community oriented, taking part in local events and giving back to a number of charities. The community has, in turn, embraced the shop. Locals feel welcomed and appreciated during every visit. Many patrons are on a first-name basis with the staff and are more than just repeat customers.

Kelly and Constantine Mihalis, owners of Beer Growler Nation in Decatur. *Ron Smith.*

Crafty Draught

Andre Airich found himself spending too much money on craft beer at a local chain sports bar. So, he decided to go into business selling the craft beer he loves. After a location was secured in a high-traffic area of Cumming (almost due north of Downtown Atlanta) and Andre waited for a county ordinance change, Crafty Draught was born. The business was Forsyth County's first growler shop, and people are getting used to the concept. "People still come in and think it's a bar," says Airich. Not even samples are allowed to be consumed on site under the county ordinance. However, the twenty taps are kept busy filling growlers for patrons to take home.

The space is unique for a growler shop. Crafty Draught has vibrant colors and an upbeat vibe, not to mention ping-pong and a corn hole toss for visitors. A second store opened in May 2013 in Alpharetta. Andre and his crew bring youthful exuberance and extensive knowledge of their product to the stores, creating a lively yet friendly atmosphere.

Celebrating Beer, Atlanta Style

Beer festivals have come a long way since St. Patrick's Day and Oktoberfest bashes. Atlanta now hosts a beer festival for every reason and every season. Nearly every town in the Atlanta environs can boast of a local beer fest. One is even named for infamous Atlanta saloon resident John Henry "Doc" Holliday.

Among the multitude of events, one stands out as particularly unique in its approach—Hotoberfest. The creation of Alan Raines and Tryon Rosser, Hotoberfest focuses on their shared loves of earth, beer and music. Sustaining natural resources is one of the objectives of the festival. The funds raised are donated to nonprofit groups that support ecological sustainability goals. Hotoberfest enlists local vendors and recycles every possible resource used in its celebration.

Raines and Rosser recognize that a beer celebration would fall short without top-notch beer on hand. Each year, they procure several pallets of small wooden casks that have been used to mature bourbon, rye whiskey or brandy. The casks are given to breweries and brewpubs to produce unique

Alan Raines and Tryon Rosser standing next to empty distillery barrels that will be used by breweries and brewpubs to age beer for the annual Hotoberfest Beer Festival. *Ron Smith.*

barrel-aged beers for the event. The festival offers not only these wood-aged beers but also other rare brews, giving casual drinkers and enthusiasts alike the opportunity to taste one-of-a-kind beers.

Attendees of Hotoberfest are given the opportunity to locate and rate their favorite festival beers in an electronic database that is available for smart phones, tablets and laptops. This voting system driven by attendees makes Hotoberfest the nation's largest consumer-judged craft beer competition.

LIKE A PHOENIX RISING FROM THE FLAMES

In tracing beer's recorded history in Atlanta, we indirectly witnessed the evolution of beer from its origins as a noncontroversial element of society, through its status as a scourge during Prohibition, into its uncertain future after repeal. The first half of the twentieth century was a difficult time for hometown beer, including at least two periods when no locally produced beer was commercially available.

What we see now in the Atlanta area would likely not have been predicted twenty years ago by the adventurers trying to reintroduce the local brewery. We mourn the loss of a few breweries that were a bit ahead of their time and did not make it to the beer resurgence. But let's also celebrate the entrepreneurs who persevered to establish the new generation of local breweries and brewpubs.

In history, brewing created demand for coopers, blacksmiths, wagon makers and coal producers. Modern brewing similarly promotes other trades: transportation and distribution, draft line cleaning services, marketing, hop farming, grain production and malting, packaging (glass, cardboard and aluminum) and retail. Perhaps we should also mention the staggering income from federal, state and local taxes and licensing.

By generating interest in craft beer, national and local brewers have paved the way for new beer-centered businesses. *Beer Connoisseur* magazine (with related online service) is an Atlanta-based business appealing to the more savvy beer lover. In 2009, the first issue of *Beer Connoisseur* hit Atlanta shelves as a high-end periodical with reviews, industry interviews and gorgeous glossy photos. The *Atlanta Journal-Constitution* has a long-running beer section in which journalist Bob Townsend regularly reviews people and places in the local beer scene. The city is also home to countless bloggers, reviewers and bulletin boards. Beer has recently popped up in more surprising formats:

Frozen Pints and Happy Hour Confections are putting craft beer to use as the basis for ice cream, waffles, cupcakes and other treats.

Despite this progress, it is our sense that Atlanta is still finding its beer identity. The metropolitan area is vast with distinct communities demonstrating local culture; perhaps beer establishments will uniquely support the communities in which they are based. Such diversity might become the hallmark of Atlanta's maturity in the beer realm. Ironically, this evolution brings the story full circle to where beer began in America's cities—local shops serving nearby residents and people carrying their growlers down the street to fetch fresh, local beer. Many of today's brewpubs, package stores and growler shops are geographically positioned to serve their communities in this way.

We believe that in addition to local beer enthusiasts being better served, energy is building to establish Atlanta as a destination location for those interested in craft beer in all its forms. Visibility of Atlanta as an award-winning beer town is increasing as local brewers collaborate with brewers from other states and even outside the country. Some Atlanta breweries are also expanding their distribution areas farther outside the state.

By embracing an outward-reaching and progressive beer culture, Atlantans can capitalize on the trend toward beer tourism, promoting trade through both direct sales and indirect supporting services. The economic benefit creates a winning situation for both those interested in beer and those who are not. Local brewers, pub owners and beer enthusiasts look forward—with bottles, pint glasses and chalices raised high—to this promising future.

BIBLIOGRAPHY

BOOKS AND JOURNALS

Ade, George. *The Old-Time Saloon: Not Wet, Not Dry, Just History.* New York: Ray Long & Richard Smith, Inc., 1931.

Ansley, J.J. *History of the Georgia Woman's Christian Temperance Union: From Its Organization 1883 to 1907.* Columbus, GA: Gilbert Printing Company, 1914.

Atlanta History Center. *Atlanta City Brewing Company Minute Book, 1876–1905.* Atlanta, GA: Gift of Abrams Books, Inc., 1968.

Baab, Bill. *Augusta on Glass.* Augusta, GA: self-published, 2007.

Ball, Mays. "Prohibition in Georgia: Its Failure to Prevent Drinking in Atlanta and Other Cities." *Putnam's Monthly and the Reader, A Magazine of Literature, Art and Life* 5 (October 1908–March 1909).

Baron, Stanley. *Brewed in America: A History of Beer and Ale in the United States.* Boston: Little, Brown, and Company, 1962.

Bauerlein, Mark. *Negrophobia: A Race Riot in Atlanta, 1906.* San Francisco: Encounter Books, 2001.

Blass, Kimberly, and Michael Rose. *Atlanta Scenes: Photojournalism in the Atlanta History Center Collection (Images of America).* Mount Pleasant, SC: Arcadia Publishing, 1998.

Buffington, Perry, and Kim Underwood. *Archival Atlanta: Forgotten Facts and Well-kept Secrets from Our City's Past.* Atlanta, GA: Peachtree Publishers, Ltd., 1996.

Burns, Rebecca. *Rage in the Gate City: The Story of the 1906 Atlanta Race Riot.* Cincinnati, OH: Emmis Books, 2006.

Coker, Joe. *Liquor in the Land of the Lost Cause: Southern White Evangelicals and the Prohibition Movement.* Lexington: University Press of Kentucky, 2007.

Davis, Angela. *Blues Legacies and Black Feminism: Gertrude Ma Rainey, Bessie Smith and Billie Holiday.* New York: Vintage Books, 1998.

Davis, Marni. *Jews and Booze: Becoming American in the Age of Prohibition.* New York: New York University Press, 2012.

Dorsey, Allison. *To Build Our Lives Together: Community Formation in Black Atlanta, 1875–1906.* Athens: University of Georgia Press, 2004.

Eckhardt, Fred. *The Essentials of Beer Style: A Catalog of Classic Beer Styles for Brewers and Beer Enthusiasts.* Portland, OR: Fred Eckhardt Communications, 1989.

Endolyn, Osayi. "Sweet Dreams: An Oral History of SweetWater." *Atlanta Magazine* (June 2013).

Fennessy, Steve. "Beer Bust: How State Laws Leave Georgia Microbreweries with a Hangover." *Creative Loafing* (October 2004).

Garrett, Franklin. *Atlanta and Environs: A Chronicle of Its People and Events.* Vols. 1–2. Athens: University of Georgia Press, 1969.

———. *Yesterday's Atlanta.* Miami, FL: E.A. Seemann Publishing, Inc., 1974.

"Glee Club Tours Continent by Plane." *Cornell Alumni News* 56, no. 15 (1954).

Godshalk, David. *Veiled Visions: The 1906 Atlanta Race Riot and the Reshaping of American Race Relations.* Chapel Hill: University of North Carolina Press, 2005.

Hanleiter, William. *Atlanta City Directory 1871.* Atlanta, GA: William R. Hanleiter Publisher, 1871.

Hertzberg, Steven. *Strangers in the Gate City: The Jews of Atlanta, 1845–1915.* Philadelphia: Jewish Publication Society of America, 1978.

Jackson, Michael. *The World Guide to Beer: The Brewing Styles, the Brands, the Countries.* Englewood Cliffs, NJ: Prentice Hall Press, 1977.

Johnson, Sharon Peregrine, and Byron A. Johnson. *The Authentic Guide to Drinks of the Civil War Era, 1853–1873.* Gettysburg, PA: Thomas Publications, 1992.

Kislingbury, Rodger. *Saloons, Bars & Cigar Stores: Historical Interior Photographs.* Pasadena, CA: Waldo and Van Winkle Publishers, 1999.

Kuhn, Clifford, Harlon Joye and Bernard West. *Living Atlanta: An Oral History of the City, 1914–1948.* Athens: University of Georgia Press, 1990.

Mahan, Mary Jane. *Love at the Pub: An Insider's Guide to Craftsmanship, Conversation, and Community at the Brick Store Pub.* Bloomington, IN: iUniverse, 2009.

Mittelman, Amy. *Brewing Battles: A History of American Beer.* New York: Algora Publishing, 2007.

Moore, Hammond. *The Negro and Prohibition in Atlanta, 1885–1887.* (Reprinted) *South Atlantic Quarterly* 69, no. 1 (Winter 1970).

Murdock, Catherine. *Domesticating Drink: Women, Men, and Alcohol in America, 1870–1940.* Baltimore, MD: Johns Hopkins University Press, 1998.

Papazian, Charlie. *The Complete Joy of Homebrewing.* 3rd edition. New York: Harper Collins Publishers, Inc., 2003.

Pine, Bob. "Pabst: The Brewery that Overcame Every Adversity." *American Breweriana Journal* 66 (January–February 1994).

Pioneer Citizens' History of Atlanta, 1833–1902. Atlanta, GA: Byrd Printing Company, 1902.

Pruett, Jed. "The Contested Gate City: Southern Progressivism's Roots in Atlanta's Local Politics, 1885–1889." Honors thesis, University of Tennessee, 2011.

Roberts, Gary. *Doc Holliday: The Life and Legend.* Hoboken, NJ: John Wiley & Sons, Inc., 2006.

Salter, Lillian. *Carling Brewery.* A small typed article from the files of the Atlanta History Center. Atlanta, GA: Atlanta History Center, 1984.

Scomp, Henry. *King Alcohol in the Realm of King Cotton...* Chicago: Blakely Print Company, 1888.

Sharpe, Kip. *The Atlanta Brewery.* A small typed article from the files of the Atlanta History Center. Atlanta, GA: Atlanta History Center, circa 1980s.

Shavin, Norman. *Underground Atlanta.* Atlanta, GA: Capricorn Corporation, 1973.

Shavin, Norman, and Bruce Galphin. *Atlanta: Triumph of a People; An Illustrated History.* Atlanta, GA: Capricorn Corporation, 1982.

Sholes, A.E. *Sholes' Directory of the City of Atlanta.* Vol. 2. Atlanta, GA: A.E. Sholes Publisher, 1878.

Sismondo, Christine. *America Walks into a Bar: A Spirited History of Taverns and Saloons, Speakeasies and Grog Shops.* New York: Oxford University Press, Inc., 2011.

Small, Sam. *Pleas for Prohibition.* Atlanta, GA: printed for the author, 1890.

Smith, George. *The Story of Georgia and the Georgia People, 1732 to 1860.* Macon, GA: Franklin Printing and Publishing Company, 1900.

Southeastern Brewing Co. v. Blackwell, Secretary of State of South Carolina, et al. 80 F.2d 607 No. 3917 (4th Cir., E.D. S.C. 1935).

"Southern Brewery Quits, Blames War Conditions." *Billboard* 56, no. 2 (January 8, 1944): 64.

Thornton, Mark. *The Economics of Prohibition.* Salt Lake City: University of Utah Press, 1991.

Tremblay, Victor, and Carol Horton Tremblay. *The U.S. Brewing Industry: Data and Economic Analysis.* Cambridge: Massachusetts Institute of Technology, 2005.

Van Wieren, Dale. *American Breweries II.* West Point, NY: East Coast Breweriana Association, 2005.

Virginia Writers' Project. *Pigsfoot Jelly & Persimmon Beer: Foodways from the Virginia Writers' Project.* Edited by Charles L. Perdue Jr. Santa Fe, NM: Ancient City Press, 1992.

Wahl, Robert, and Arnold Wahl. *American Brewers Review* 12 (June/July 1898–1899).

————. *American Brewers Review* 21 (July 1907).

Williams, Paige. "Cheers: A Tour of Atlanta's Award-Winning Brewpubs and Micro-Breweries." *Atlanta Magazine* (May 2003).

Williford, William. *Peachtree Street, Atlanta.* Athens: University of Georgia Press, 1962.

Wilson, John. *Atlanta As It Is: Being a Brief Sketch of Its Early Settlers, Growth, Society...* New York: Little, Rennie & Company, 1871.

Woolsey, David. *Libations of the Eighteenth Century: A Concise Manual for the Brewing of Authentic Beverages from the Colonial Era of America and of Times Past.* Boca Raton, FL: Universal Publishers, Inc., 2002.

World Almanac and Encyclopedia, 1908. New York: Press Publishing Company, 1908.

Writers' Program of the Work Projects Administration in the State of Georgia. *Atlanta: A City of the Modern South.* New York: Smith & Durrell, 1942.

Zimmerman, Jonathan. *Distilling Democracy: Alcohol Education in America's Public Schools, 1880–1925.* Lawrence: University Press of Kansas, 1999.

Newspaper Articles

Atlanta Journal. "Carling Brewery Opens Plant Here." June 17, 1958.

Boston Herald. "Scofflaw Article." January 16, 1924.

(New London, CT) Day. "Atlanta's Political Junction." August 9, 1986.

(Lexington, NC) Dispatch. "Schweitzer-ized Beer Made from Secret Swiss Formula." June 25, 1949.

Fort Wayne (IN) Sentinel. "Good Use for Brewery." December 14, 1907.

New York Times. "Microbrewers Set the Pace, and Draw the Heat." November 20, 1994.

BIBLIOGRAPHY

NEWSPAPERS SEARCHED IN ELECTRONIC FORMAT

(Atlanta) Constitution, 1868
Atlanta Constitution, 1869–2001
Atlanta Daily Herald, 1873–76
Atlanta Georgian, 1906–11
Atlanta Intelligencer, 1851, 1854–71
Atlanta Journal, 1883–2001
Atlanta Journal-Constitution, 2001–present
Atlantian, 1911–22
Georgia Literary and Temperance Crusader, 1860–61
Southern Confederacy, 1861–64
Sunny South, 1875–1907
Weekly Constitution, 1869–1882

INTERVIEWS AND PERSONAL DISCUSSIONS

Airich, Andre, coowner of Crafty Draught. May 2013.

Amerson, Romie, general manager of Augustine's. May 2013.

Baab, Bill, author and expert on antique bottles. September 2012.

Baker, Jonathan, marketing for Monday Night Brewery. March 2013.

Bensch, Freddy, owner of SweetWater Brewing Company. E-mail communication with Ron Smith. May 2013.

Blanchard, Dave, coowner of the Brick Store Pub. April 2013.

Bradley, Jonny, brewer at Cherry Street Brewing Cooperative. May 2013.

Budd, Robert, president of Red Brick Brewing Company. March 2013.

Comer, Colin, coowner of the Marlay House. April 2013.

Cowan, Nathan, coowner of Eventide Brewing. June 2013.

Crudder, Fred, beer manager for Taco Mac. May 2013.

Dempsey, Randy, owner of and brewer at O'Dempsey's Beer to Die For. May 2013.

Engleman, Neal, assistant brewer at Wrecking Bar Brewpub. April 2013.

Evans, Doug, coowner of and brewer at Strawn Brewing Company. April 2013.

Fowler, Nick, production engineer at Red Brick Brewing Company. March 2013.

Golden, Glenn, owner of and brewer at Jailhouse Brewing Company. March 2013.

Gunn, Molly, coowner of the Porter Beer Bar. April 2013.

Hager, Harry, beer expert at Beer Growler Nation, May 2013.

Harris, Happy, craft manager at General Wholesale Beer Company (formerly, brewer at John Harvard's). April 2013.

Hedeen, Scott, owner of and brewer at the Burnt Hickory Brewery. March 2013.

Hoglund, Dan, Atlantic Company breweriana collector. Telephone interview with Ron Smith. March 2013

Holley, Eddie, coowner of Ale Yeah! Craft Beer Market. February 2013.

Hubbard, Doug, brewer at Marthasville Brewing Company. Telephone interview with Ron Smith. March 2013.

Jones, Ken, Atlantic Company expert and breweriana collector. Telephone and in-person interviews with Ron Smith. January and March 2013.

Karakos, George, and Martha Karakos, former operators of Atkins Park Tavern. May 2013.

Lamb, George, partner at Marthasville Brewing Company. Telephone interview with Ron Smith. March 2013.

Lubrant, Richard, charter member of Covert Hops Society. Telephone interview with Ron Smith. May 2013.

Maloof, Brian, owner of Manuel's Tavern. March 2013.

Mihalis, Constantine, and Kelly Mihalis, owners of Beer Growler Nation. April 2013.

Mitchell, Tad, coowner of Six Feet Under Pub & Fish House and breweriana collector. February 2013.

Moran, Crawford, brewmaster at 5 Seasons Restaurant and Brewery (former owner of and brew master at Dogwood Brewing Company). April 2013.

Raines, Alan, cofounder of Hotoberfest. May 2013.

Roberts, John "JR," partner and brew master at Max Lager's Wood-Fired Grill & Brewery. March 2013.

Rodriguez, Anthony, retail and promotions manager at Red Brick Brewing Company. March 2013.

Rosser, Tryon, cofounder of Hotoberfest. May 2013.

Sandage, Bob, coowner of and brew master at Wrecking Bar Brewpub. April 2013.

Scoggins, Dow, former owner of Friends Brewing Company and Helenboch Brewery. Telephone interview with Ron Smith. May 2013.

Strawn, Will, coowner of and brewer at Strawn Brewing Company. April 2013.

Tanner, Nick, founder of and brewer at Cherry Street Brewing Cooperative. May 2013.

Thomas, Bobby, coowner of and brew master at Red Hare Brewing Company. April 2013.

Torres, Kraig, owner of Hop City. May 2013.

BIBLIOGRAPHY

VISUAL MEDIA

Vintage TV Beer Commercials: 100 Beer Commercials of the 1950s and '60s. Cleveland: Schnitzelbank Press, 2003.

Ward, Geoffrey. *Prohibition.* Directed by Ken Burns and Lynn Novick. Hollywood: Prohibition Film Project, Inc., 2011.

WEBSITES

American Breweriana Association. http://www.americanbreweriana.org (accessed April 26, 2013).

American Brewery History. BeerHistory.com. http://www.beerhistory.com/library/ (accessed July 28, 2012).

American Homebrewers Association. http://homebrewersassociation.com (accessed January 6, 2013).

"Antique Beer Bottles." Antique Bottles Collector's Haven. http://www.antiquebottles.com/beer (accessed July 31, 2012).

"Antique Beer Bottles." Collectors Weekly. http://www.collectorsweekly.com/bottles/beer (accessed July 31, 2012).

Atkins Park Taverns and Restaurants. http://www.atkinspark.com (accessed August 7, 2012).

Atlanta Beer Guide. Beer Info.com. http://www.beerinfo.com/index.php/pages/atlantabeerguide.html (accessed January 8, 2013).

Atlanta Business Chronicle. http://www.bizjournals.com/atlanta (accessed March 27, 2013).

Atlanta City Brewing Company. Beer Bottle Library. http://brucemobley.com/beerbottlelibrary/ga/atlanta/atlanta.htm (accessed July 31, 2012).

Atlanta City Directories, 1878–1922. Located at Digital Library of Georgia. http://dlg.galileo.usg.edu/cgi-bin/meta.cgi (accessed July 20, 2012).

"Atlanta Old and New: Prehistory to 1847." Roadside Georgia. http://roadsidegeorgia.com/city/atlanta01.html (accessed July 29, 2012).

"Atlanta Timeline: A Chronology of the History of Atlanta 1782–2010." http://citycouncil.atlantaga.gov/historynew.htm (accessed March 26, 2013).

Atlanta Time Machine. www.atlantatimemachine.com (accessed July 28, 2012).

"Atlantic Beer." RustyCans.com. http://www.rustycans.com/COM/month0504.html (accessed July 31, 2012).

Belgian Brewers. http://www.beerparadise.be (accessed July 22, 2012).

Bieretikettenkatalog der USA (Beer Label Catalog of the USA). http://www.klausehm.de/Ubersichtusa1.html (accessed July 22, 2012).

BIBLIOGRAPHY

Blocker, Dr. Jack S. "Did Prohibition Really Work? Alcohol Prohibition as a Public Health Innovation." *American Journal of Public Health* 96, no. 2. http://www.ncbi.nlm.nih.gov/pmc/articles/PMC1470475 (accessed November 24, 2012).

Bob Kay Beer Labels. http://home.comcast.net/~beerlabel/V2newfinds.htm (accessed January 3, 2013).

Brewers Association. Craft Brewing Statistics. Number of Breweries. http://www.brewersassociation.org/pages/business-tools/craft-brewing-statistics/number-of-breweries (accessed February 7, 2013).

"Brewing During the Civil War Era." Professor Good Ales. http://professorgoodales.net/archives/4611 (accessed July 30, 2012).

Confederate Railroads. http://www.csa-railroads.com (accessed August 2, 2012).

Der Biergarten. http://www.derbiergarten.com (accessed April 20, 2013).

Digital Collections. Georgia State University. http://digitalcollections.library/gsu.edu/cdm (accessed August 8, 2012).

Dyer, Candice. "The Royal Peacock." Antics in Candyland. http://anticsincandyland.wordpress.com/2010/05/11/the-royal-peacock-dont-you-wish-you-were-there-back-then-at-atlantas-club-beautiful/ (accessed August 7, 2012).

Eventide Brewing. http://eventidebrewing.com (accessed June 6, 2013).

5 Seasons Brewing Company. http://www.5seasonsbrewing.com (accessed April 17, 2013).

Fold 3: Historical Military Records. http://www.fold3.com (accessed July 24, 2012).

Georgia Beer Coasters. Beer Coaster Mania. http://beercoastermania.com/ga/gacoasters.html (accessed December 20, 2012).

Georgia. Beer Expedition. www.beerexpedition.com/ga/ (accessed July 31, 2012).

Georgia Breweries. Micro Label Guide. http://ga.microlabelguide.com (accessed January 3, 2013).

"Georgia Historic Newspapers: Atlanta." Digital Library of Georgia. http://atlnewspapers.galileo.usg.edu (accessed July 20, 2012).

German Beer Institute: German Beer Portal for North America. http://www.germanbeerinstitute.com (accessed July 22, 2012).

Glass Discoveries. http://www.glassdiscoveries.com (accessed July 31, 2012).

Hey Mabel Black Label! http://heymabelblacklabel.com (accessed July 22, 2012).

"History of Buckhead." Buckhead Heritage Society. http://www.buckheadheritage.com/node/39 (accessed July 28, 2012).

Hutchbook.com. http://www.hutchbook.com/default.html (accessed July 31, 2012).

Jones, Ken. "Bock Beers of the Atlantic Company." *Atlantic Waves* (October 2006). http://www.bccaatlantic.com/Newsletters/Oct2006.pdf (accessed August 4, 2012).

———. "Collecting Atlantic Company Labels: Part 2." *Atlantic Waves* (June 2004). http://www.bccaatlantic.com/Newsletters/June2004.pdf (accessed August 4, 2012).

Lehfeldt, Martin. "Manuel's Tavern: A Historic Stop for Visiting Historians." From the 2007 supplement to the 121st annual meeting of Atlanta and Historians. American Historical Association. http://www.historians.org/perspectives/issues/2006/0612/07AMSupplement/07AMSup22.cfm (accessed March 18, 2013).

Lost Beers. http://www.thelostbeers.com (accessed February 14, 2013).

Marietta Street Artery Association: Architectural History. http://www.artery.org/08_history/2-Atlanta-GA.htm (accessed August 5, 2012).

Michael Jackson's Beer Hunter. http://www.beerhunter.com (accessed August 11, 2012).

Midway Pub. http://www.themidwaypub.com (accessed June 9, 2013).

Moe's and Joe's: Atlanta. http://www.moesandjoes.com/Welcome.html (accessed April 2, 2013).

"Mountain View, GA." Wikipedia. en.wikipedia.org/wiki/Mountain_View_Georgia (accessed July 22, 2012).

Park Tavern. http://www.parktavern.com (accessed April 2, 2013).

Philp, Matt. "Jews & Prohibition Booze." Wine-Searcher (September 2012). http://www.wine-searcher.com/m/2012/09/the-clash-of-boozy-jews-and-abstaining-protestants (accessed November 23, 2012).

Prohibition Party. http://www.prohibitionparty.org (accessed November 20, 2012).

"Prohibition Party." Wikipedia. http://en.wikipedia.org/wiki/Prohibition_Party (accessed November 20, 2012).

Real Beer. http://www.realbeer.com/library (accessed July 30, 2012).

"Red Brick Ale." *All About Beer* 23, no. 5 (November 2002). http://allaboutbeer.com/learn-beer/reviews/beer-talk/2002/11/red-brick-ale/ (accessed March 26, 2013).

"Royal Peacock Club: Atlanta, Georgia." Discover Black Heritage. http://discoverblackheritage.com/royal-peacock-club (accessed August 7, 2012).

Saloon Tokens. http://www.saloontokens.info (accessed October 8, 2012).

Sanborn® Fire Insurance Maps for Atlanta, 1886–1922. Digital Library of Georgia. http://dlg.galileo.usg.edu/sanborn (accessed November 20, 2012).

"Short History of Atlanta, A, 1782–1859." City-Directory, Inc. http://www.city-directory.com/atlanta/history (accessed July 22, 2012).

Soda & Beer Bottles of North America. http://www.sodasandbeers.com (accessed July 31, 2012).

Southern Antique Bottles. http://www.southernbottles.com (accessed July 31, 2012).

"Streetcars in Atlanta." Georgia's Railroad History and Heritage. http://railga.com/oddend/streetrail/atlantastr.html (accessed August 21, 2012).

"Tabernacle, the." Wikipedia. en.wikipedia.org/wiki/The_Tabernacle (accessed July 28, 2012).

Tavern Trove.com. http://www.taverntrove.com/whatsnew.php (accessed October 30, 2012).

"Temperance Movement." New Georgia Encyclopedia. http://www.georgiaencyclopedia.org/nge/Article.jsp?id=h-828 (accessed November 21, 2012).

Traetto, Lauren. "Manuel's Tavern Lives: A Modern Folk Tale." *Purge* (August 2012). http://purgeatl.com/2012/08/01/manuels-tavern-lives-a-modern-folk-tale (accessed March 22, 2013).

"Underground Atlanta." Wikipedia. en.wikipedia.org/wiki/Underground_Atlanta (accessed August 8, 2012).

Underwood, Madison. "ABC Board attorney: Birmingham Homebrewing equipment seizure was not a 'raid.'" *AL.com*, September 21, 2012. http://blog.al.com/spotnews/2012/09/abc_board_attorney_brewing_equ.html (accessed April 10, 2013).

"Vintage Beer Cans." Collectors Weekly. http://www.collectorsweekly.com/breweriana/cans (accessed July 31, 2012).

Waterhouse, Jon. "Ghosts of Hotspots Past." *Creative Loafing*. http://clatl.com/atlanta/ghosts-of-hotspots-past/Content?oid=1241280 (accessed March 18, 2013).

"West End." City of Atlanta. http://www.atlantaga.gov/index.aspx?page=467 (accessed July 20, 2012).

Wrecking Bar Brewpub and the Marianna. http://www.wreckingbarbrewpub.com (accessed April 5, 2013).

INDEX

ABOUT THE AUTHORS

RONALD J. SMITH

Dubbed "the beer guru" after speaking at a Brockett Pub House and Grill's beer tasting, Ron has been interested in artisanal beer since the early 1990s and studies the subject extensively. He has been fortunate to travel all over the United States and parts of Europe to try locally produced brews. Ron has hosted beer and food dinners, beer tastings and beer education sessions. An obsessive researcher, he is particularly fond of investigating historical breweries in the southeastern United States. An occasional homebrewer, his favorite beer styles are sour ales and saisons. He is a member of the Georgia Craft Brewers' Guild and the American Breweriana Association.

Photo by Judy Kuniansky.

MARY O. BOYLE

A detailed researcher and hawk-eyed editor, Mary is no stranger to writing or craft beer. Cohost of several beer dinners and a craft beer enthusiast since 2003, Mary currently is webmaster at www.beerguruatl.com, produces a local pub's newsletter and is adding a line of beer-focused jewelry to her handcrafted jewelry business. As with her coauthor, Mary has traveled to many locations and sampled the local offerings. Her favorite beer style is Belgian quadrupel. After adopting a gluten-free lifestyle, she has explored gluten-free alternatives (and truly appreciates 5 Seasons Westside's keeping their Elmer's gluten-free beer on tap).

CPSIA information can be obtained
at www.ICGtesting.com
Printed in the USA
LVHW080741301219
642040LV00002B/11/P